Advances in Intelligent Systems and Computing

Volume 431

Series editor

Janusz Kacprzyk, Polish Academy of Sciences, Warsaw, Poland
e-mail: kacprzyk@ibspan.waw.pl

AF786304

About this Series

The series "Advances in Intelligent Systems and Computing" contains publications on theory, applications, and design methods of Intelligent Systems and Intelligent Computing. Virtually all disciplines such as engineering, natural sciences, computer and information science, ICT, economics, business, e-commerce, environment, healthcare, life science are covered. The list of topics spans all the areas of modern intelligent systems and computing.

The publications within "Advances in Intelligent Systems and Computing" are primarily textbooks and proceedings of important conferences, symposia and congresses. They cover significant recent developments in the field, both of a foundational and applicable character. An important characteristic feature of the series is the short publication time and world-wide distribution. This permits a rapid and broad dissemination of research results.

Advisory Board

Chairman

Nikhil R. Pal, Indian Statistical Institute, Kolkata, India
e-mail: nikhil@isical.ac.in

Members

Rafael Bello, Universidad Central "Marta Abreu" de Las Villas, Santa Clara, Cuba
e-mail: rbellop@uclv.edu.cu

Emilio S. Corchado, University of Salamanca, Salamanca, Spain
e-mail: escorchado@usal.es

Hani Hagras, University of Essex, Colchester, UK
e-mail: hani@essex.ac.uk

László T. Kóczy, Széchenyi István University, Győr, Hungary
e-mail: koczy@sze.hu

Vladik Kreinovich, University of Texas at El Paso, El Paso, USA
e-mail: vladik@utep.edu

Chin-Teng Lin, National Chiao Tung University, Hsinchu, Taiwan
e-mail: ctlin@mail.nctu.edu.tw

Jie Lu, University of Technology, Sydney, Australia
e-mail: Jie.Lu@uts.edu.au

Patricia Melin, Tijuana Institute of Technology, Tijuana, Mexico
e-mail: epmelin@hafsamx.org

Nadia Nedjah, State University of Rio de Janeiro, Rio de Janeiro, Brazil
e-mail: nadia@eng.uerj.br

Ngoc Thanh Nguyen, Wroclaw University of Technology, Wroclaw, Poland
e-mail: Ngoc-Thanh.Nguyen@pwr.edu.pl

Jun Wang, The Chinese University of Hong Kong, Shatin, Hong Kong
e-mail: jwang@mae.cuhk.edu.hk

More information about this series at http://www.springer.com/series/11156

Jerzy Świątek · Leszek Borzemski
Adam Grzech · Zofia Wilimowska
Editors

Information Systems Architecture and Technology: Proceedings of 36th International Conference on Information Systems Architecture and Technology – ISAT 2015 – Part III

Springer

Editors
Jerzy Świątek
Faculty of Computer Science
and Management
Wrocław University of Technology
Wrocław
Poland

Adam Grzech
Faculty of Computer Science
and Management
Wrocław University of Technology
Wrocław
Poland

Leszek Borzemski
Faculty of Computer Science
and Management
Wrocław University of Technology
Wrocław
Poland

Zofia Wilimowska
Faculty of Computer Science
and Management
Wrocław University of Technology
Wrocław
Poland

ISSN 2194-5357 ISSN 2194-5365 (electronic)
Advances in Intelligent Systems and Computing
ISBN 978-3-319-28562-7 ISBN 978-3-319-28564-1 (eBook)
DOI 10.1007/978-3-319-28564-1

Library of Congress Control Number: 2016930056

© Springer International Publishing Switzerland 2016
This work is subject to copyright. All rights are reserved by the Publisher, whether the whole or part of the material is concerned, specifically the rights of translation, reprinting, reuse of illustrations, recitation, broadcasting, reproduction on microfilms or in any other physical way, and transmission or information storage and retrieval, electronic adaptation, computer software, or by similar or dissimilar methodology now known or hereafter developed.
The use of general descriptive names, registered names, trademarks, service marks, etc. in this publication does not imply, even in the absence of a specific statement, that such names are exempt from the relevant protective laws and regulations and therefore free for general use.
The publisher, the authors and the editors are safe to assume that the advice and information in this book are believed to be true and accurate at the date of publication. Neither the publisher nor the authors or the editors give a warranty, express or implied, with respect to the material contained herein or for any errors or omissions that may have been made.

Printed on acid-free paper

This Springer imprint is published by SpringerNature
The registered company is Springer International Publishing AG Switzerland

Preface

This four volume set of books includes the proceedings of the 2015 36th International Conference Information Systems Architecture and Technology (ISAT), or ISAT 2015 for short, held on September 20–22, 2015, in Karpacz, Poland. The conference was organized by the Department of Computer Science and Department of Management Systems, Faculty of Computer Science and Management, Wrocław University of Technology, Poland.

The International Conference Information Systems Architecture is organized by the Wrocław University of Technology from the seventies of the last century. The purpose of the ISAT is to discuss a state of the art of information systems concepts and applications as well as architectures and technologies supporting contemporary information systems. The aim is also to consider an impact of knowledge, information, computing, and communication technologies on managing the organization scope of functionality as well as on enterprise information systems design, implementation, and maintenance processes taking into account various methodological, technological, and technical aspects. It is also devoted to information systems concepts and applications supporting exchange of goods and services by using different business models and exploiting opportunities offered by Internet-based electronic business and commerce solutions.

ISAT is a forum for specific disciplinary research, as well as on multi-disciplinary studies to present original contributions and to discuss different subjects of today's information systems planning, designing, development, and implementation. The event is addressed to the scientific community, people involved in variety of topics related to information, management, computer, and communication systems, and people involved in the development of business information systems and business computer applications.

This year, we received 130 papers from 17 countries. The papers included in the four proceeding volumes published by Springer have been subject to a thoroughgoing review process by highly qualified peer reviewers. Each paper was reviewed by at least two members of Program Committee or Board of Reviewers. Only 74 best papers were selected for oral presentation and publication in the

36th International Conference Information Systems Architecture and Technology 2015 proceedings. The final acceptance rate was 57 %.

Professor Peter Nelsen (Denmark) presented his keynote speech on Some Insights from Big Data Research Projects. He also organized the special session on the advances in methods for managing complex planning environments.

The conference proceedings are divided into four volumes and present papers in the areas of managing complex planning environments, systems analysis and modeling, finance, logistics and market, artificial intelligence, knowledge-based management, Web systems, computer networks and distributed computing, high performance computing, cloud computing, multi-agent systems, Internet of Things, mobile systems, service-oriented architecture systems, knowledge discovery, and data mining.

We would like to thank the Program Committee and external reviewers, essential for reviewing the papers to ensure a high standard of the ISAT 2015 conference and the proceedings. We thank the authors, presenters, and participants of ISAT 2015; without them, the conference could not have taken place. Finally, we thank the organizing team for the efforts this and previous years in bringing the conference to a successful conclusion.

September 2015

Jerzy Świątek
Leszek Borzemski
Adam Grzech
Zofia Wilimowska

ISAT 2015 Conference Organization

General Chair

Leszek Borzemski, Poland

Program Co-chairs

Leszek Borzemski, Poland
Adam Grzech, Poland
Jerzy Świątek, Poland
Zofia Wilimowska, Poland

Local Organizing Committee

Leszek Borzemski, Chair
Zofia Wilimowska, Vice-Chair
Mariusz Fraś, Conference Secretary and ISAT 2015 Website Administrator
Arkadiusz Górski, Katarzyna Gwóźdź, Technical Editors
Anna Kiłyk, Ziemowit Nowak, Agnieszka Parkitna, Technical Chairmen

International Program Committee

Witold Abramowicz, Poland
Dhiya Al-Jumeily, UK
Iosif Androulidakis, Greece
Patricia Anthony, New Zeland
Zbigniew Banaszak, Poland
Elena N. Benderskaya, Russia
Leszek Borzemski, Poland
Janos Botzheim, Japan

Patrice Boursier, France
Wojciech Cellary, Poland
Haruna Chiroma, Malaysia
Edward Chlebus, Poland
Gloria Cerasela Crisan, Romania
Marilia Curado, Portugal
Czesław Daniłowicz, Poland
Zhaohong Deng, China
Małgorzata Dolińska, Poland
El-Sayed M. El-Alfy, Saudi Arabia
Naoki Fukuta, Japan
Piotr Gawkowski, Poland
Manuel Graña, Spain
Wiesław M. Grudzewski, Poland
Adam Grzech, Poland
Irena Hejduk, Poland
Katsuhiro Honda, Japan
Marian Hopej, Poland
Zbigniew Huzar, Poland
Natthakan Iam-On, Thailand
Biju Issac, UK
Arun Iyengar, USA
Jürgen Jasperneite, Germany
Janusz Kacprzyk, Poland
Henryk Kaproń, Poland
Yannis L. Karnavas, Greece
Ryszard Knosala, Poland
Zdzisław Kowalczuk, Poland
Binod Kumar, India
Jan Kwiatkowski, Poland
Antonio Latorre, Spain
Gang Li, Australia
José M. Merigó Lindahl, Chile
Jose M. Luna, Spain
Emilio Luque, Spain
Sofian Maabout, France
Zygmunt Mazur, Poland
Pedro Medeiros, Portugal
Toshiro Minami, Japan
Marian Molasy, Poland
Zbigniew Nahorski, Poland
Kazumi Nakamatsu, Japan
Peter Nielsen, Denmark
Tadashi Nomoto, Japan
Cezary Orłowski, Poland

Michele Pagano, Italy
George A. Papakostas, Greece
Zdzisław Papir, Poland
Marek Pawlak, Poland
Jan Platoš, Czech Republic
Tomasz Popławski, Poland
Edward Radosiński, Poland
Dolores I. Rexachs, Spain
José S. Reyes, Spain
Leszek Rutkowski, Poland
Gerald Schaefer, UK
Habib Shah, Malaysia
Jeng Shyang, Taiwan
Anna Sikora, Spain
Małgorzata Sterna, Poland
Janusz Stokłosa, Poland
Remo Suppi, Spain
Edward Szczerbicki, Australia
Jerzy Świątek, Poland
Eugeniusz Toczyłowski, Poland
Elpida Tzafestas, Greece
José R. Villar, Spain
Bay Vo, Vietnam
Hongzhi Wang, China
Leon S.I. Wang, Taiwan
Jan Werewka, Poland
Thomas Wielicki, USA
Zofia Wilimowska, Poland
Bernd Wolfinger, Germany
Józef Woźniak, Poland
Roman Wyrzykowski, Poland
Jaroslav Zendulka, Czech Republic
Bernard Ženko, Slovenia

ISAT 2015 Reviewers

Patricia Anthony, New Zeland
Zbigniew Banaszak, Poland
Elena N. Benderskaya, Russia
Grzegorz Bocewicz, Poland
Leszek Borzemski, Poland
Janos Botzheim, Japan
Patrice Boursier, France
Krzysztof Brzostowski, Poland

Wojciech Cellary, Poland
Gloria Cerasela Crisan, Romania
Marilia Curado, Portugal
Mariusz Czekała, Poland
Grzegorz Debita, Poland
El-Sayed M. El-Alfy, Saudi Arabia
Stefan Forlicz, Poland
Mariusz Fraś, Poland
Naoki Fukuta, Japan
Piotr Gawkowski, Poland
Manuel Graña, Spain
Dariusz Gąsior, Poland
Adam Grzech, Poland
Irena Hejduk, Poland
Katsuhiro Honda, Japan
Zbigniew Huzar, Poland
Biju Issac, UK
Jerzy Józefczyk, Poland
Krzysztof Juszczyszyn, Poland
Yannis L. Karnavas, Greece
Radosław Katarzyniak, Poland
Grzegorz Kołaczek, Poland
Zdzisław Kowalczuk, Poland
Binod Kumar, India
Jan Kwiatkowski, Poland
Antonio Latorre, Spain
Arkadiusz Liber, Poland
Wojciech Lorkiewicz, Poland
Zygmunt Mazur, Poland
Pedro Medeiros, Portugal
Izabela Nielsen, Denmark
Peter Nielsen, Denmark
Tadashi Nomoto, Japan
Cezary Orłowski, Poland
Michele Pagano, Italy
George A. Papakostas, Greece
Marek Pawlak, Poland
Jan Platoš, Czech Republic
Tomasz Popławski, Poland
Dolores I. Rexachs, Spain
José S. Reyes, Spain
Gerald Schaefer, UK
Habib Shah, Saudi Arabia
Anna Sikora, Spain
Małgorzata Sterna, Poland

Janusz Stokłosa, Poland
Remo Suppi, Spain
Edward Szczerbicki, Australia
Jerzy Świątek, Poland
Elpida Tzafestas, Greece
José R. Villar, Spain
Tomasz Walkowiak, Poland
Hongzhi Wang, China
Leon S.I. Wang, Taiwan
Adam Wasilewski, Poland
Jan Werewka, Poland
Zofia Wilimowska, Poland
Bernd Wolfinger, Germany
Józef Woźniak, Poland
Jaroslav Zendulka, Czech Republic
Maciej Zięba, Poland
Bernard Ženko, Slovenia

ISAT 2015 Keynote Speaker

Professor Peter Nielsen, Aalborg University, Aalborg, Denmark
Topic: Some Insights from Big Data Research Projects

ISAT 2015 Invited Session

Advances in Methods for Managing Complex Planning Environments
Chair: Peter Nielsen, Denmark

Contents

Part I
Mobile and SOA Systems

Personal Wi-Fi Based Indoor Localization of Mobile Devices in Active Environment

Mariusz Fraś, Krzysztof Waśko and Tomasz Wierzowiecki

Abstract The localization techniques became very popular and widely used in recent years. Popularization of mobile devices such as smartphones particularly increased the importance of wireless localization, including indoor localization. This paper deals with two issues. First, it examines the effectiveness of selected fingerprint based Wi-Fi indoor localization methods utilized in personal localization system with use of ordinary mobile phones in non-controlled active environment in the presence of dynamic interferences. Second, the new fingerprint based positioning algorithm wkNN-Bayes that combines k nearest neighbor approach with probabilistic algorithm based on Bayes theory using normal distribution to model signal strength distribution is presented.

Keywords Indoor wireless localization · RSSI fingerprinting · Probabilistic algorithm

1 Introduction

The identification of localization of considered entity such as mobile device, user, source of activity etc. has the increasing importance in the current computer science technologies and information services. Positioning data can be utilized in such areas as shopping business, booking services, local information delivery, quality of network services, social services, search services, and many more.

Satellite navigation systems and GPS modules widely installed in the mobile devices were a big step forward in the area of positioning services. The GPS system

M. Fraś (✉) · K. Waśko · T. Wierzowiecki
Wroclaw University of Technology, Wroclaw, Poland
e-mail: mariusz.fras@pwr.edu.pl

K. Waśko
e-mail: krzysztof.wasko@pwr.edu.pl

T. Wierzowiecki
e-mail: tomasz.wierzowiecki@gmail.com

© Springer International Publishing Switzerland 2016

J. Świątek et al. (eds.), *Information Systems Architecture and Technology: Proceedings of 36th International Conference on Information Systems Architecture and Technology – ISAT 2015 – Part III*, Advances in Intelligent Systems and Computing 431, DOI 10.1007/978-3-319-28564-1_1

launched in 70s by U.S. Department of Defense, consists of tens of satellites transmitting radio signal which allows the receiver to calculate its geographical position based on the measured signal propagation time. Since year 2000 the system permits to determine the position with an rms (root mean square) accuracy of 4–12 m for all civil users [1] and with 2.6–12 m rms accuracy for registered GPS users [2]. The GPS based systems such as wide-area DGPS (Differential GPS) or various Satellite Based Augmentation Systems (SBAS) allow to achieve even a centimeter accuracy [3]. However, the usability of GPS is not as big as we would like. In the cities the accuracy drops because of interference and high density of buildings. Even more important is the fact that people reside a lot of the time inside the buildings or other places being out of the GPS signal range. In large buildings (e.g. shopping centers) the positioning system is very desirable but also not easy to accomplish. The other solutions for indoor localization are based on GSM standard—e.g. in [4], RFID (Radio Frequency Identification) tags—e.g. in [5, 6], and UWB (Ultra Wideband) technology—e.g. in [7]. However the greatest potential and most promising seem to be techniques based on Wi-Fi technology.

The rapid development the local Wi-Fi networks in last years and widespread availability of mobile phones (smartphones) equipped with Wi-Fi modules allow to widely utilize Wi-Fi based indoor localization. There is a lot of works that considered various approaches, algorithms, ideas, etc. for indoor Wi-Fi localization during several recent years [8, 9]. Most of them usually presents results of experiments made in clear environment (similar to laboratory conditions), with use of strictly selected equipment, to show absolute accuracy of proposed methods and make comparisons clear.

The purpose of this work is twofold—first, to show the effectiveness of selected popular fingerprint based Wi-Fi indoor localization methods utilized in personal system with use of ordinary mobile phones in non-controlled active environment in the presence of dynamically changing interferences. As effectiveness the positioning accuracy is considered. Second, to propose a modification of classical probabilistic algorithm, that improves effectiveness of object localization—wkNN-Bayes algorithm that combines k nearest neighbor approach with probabilistic algorithm based on Bayes theory using normal distribution to model signal strength distribution. In Sect. 2 the work presents brief description and characteristics of Wi-Fi based techniques used for indoor localization. In Sect. 3 the new approach for indoor positioning is presented. In Sect. 4 the project and results of experiments are presented followed by final remarks in Sect. 5.

2 Wi-Fi Based Indoor Localization Methods

The most common solutions for indoor localization of mobile objects are methods based on detecting and processing signal of IEEE 802.11 standard (Wi-Fi) networks. The advantages of this approach is rapidly increasing number of Wi-Fi networks, popularization of smartphones equipped with Wi-Fi communication module as a standard, and API to collect Wi-Fi data available for popular smartphones system software. The most popular methods can be generally classified into two categories: direct calculations of position with use of measured signal characteristic (e.g. signal strength or signal delay) usually using lateration approach, and fingerprinting—using mapping of signal characteristic to indoor area.

The lateration (trilateriation, multirateration) technique of computing position consists in measuring distance to multiple points of known positions. The intersection of distance circles indicates object position. Having at least three measurements from three different non-linear points the trilateration permits compute localization in two dimensions. The three main approaches to measure distance are: direct measurements (difficult and very little applicable to mobile phones), measurement of signal attenuation, and measurement of time it takes to travel signal between the object and the known point.

The estimation of the distance between the device and each transmitter can be obtained on the basis of the relationship between radio signal strength (RSSI—Received Signal Strength Indication) and distance taken from theoretical model of signal propagation which express relationships between the signal power, the power of the transmitter, the wavelength, the path loss coefficient which depends on the environment, the gains of the receiver and the transmitter, and the distance between the receiver and the transmitter. However, in real world conditions, the theoretical and real signal strength differ because of such issues as reflections, refractions, the presence of other dynamic objects (e.g. people), etc. what results in poor distance estimation. In [10] was shown that indoor signal noise is definitely larger than outside the buildings. Therefore, estimation methods of distance determination must be additionally applied.

For trilateration of noised signal the LSE (Least Square Error), NLSE (Nonlinear Least Square Error), and WLSE (Weighted Least Square Error) methods and their extensions are commonly used to estimate distance—e.g. [11]. WLSE method assumes that for more distant transmitters the measured signal error is likely larger and such measurements should have less impact on calculations. Various works show various error values for tested similar solutions.

The other approach is to measure the time it takes to "fly" the signal from the object to access point (and back as a variant)—ToF (Time of Flight) parameter. In [12] the solution that uses Cristian algorithm to measure time and finally distance was presented. In the experiments the mean error level of position estimation was about 2 m. However, the measurements were performed in laboratory conditions. Apart from that the problems in such approach are time synchronization of interacting objects, the granularity of clocks and several other. Although, as shown in [13], the distance estimation accuracy can be within 1–3 m, this technique is difficult to use in diverse environment such as shopping centers and the like.

Very popular solutions for indoor localization of mobile objects are fingerprinting based methods that consist in constructing the map of values of chosen signal characteristic in chosen points of the area (fingerprint). The measured signal parameter is usually proportional to the distance between the point and the transmitter. Comparison of the measured value of this characteristic and the map permits to estimate localization. The method consists of two phases:

- offline phase—during this step the measured metrics in chosen points of the terrain are collected and recorded in database, including the localization determined manually or by another measurement,
- online phase—collecting momentary metrics and on the basis of processing using data from database determining the current localization of the object.

Usually used metrics are Received Signal Strength Indicator (RSSI), Time of Flight (ToF), and Angle of Arrival (AoA) [14].

Due to measurement inaccuracy simple comparison of the measured metric and the map usually doesn't indicate one clear position and two or more positions can fit our measurement. Hence additional processing must be performed to find the closest match from known points. For often used RSSI Fingerprinting the following approaches are mostly used:

deterministic methods—nearest neighbor based methods,
probabilistic methods—methods using signal strength distribution approximation and maximum likelihood estimation,
statistical analysis and machine learning based methods,
the methods that uses various AI approaches, e.g. neural networks.

Very popular because of their simplicity are firstly proposed nearest neighbor (NN) and k nearest neighbors (kNN) algorithms [15, 16], and their variations in later works. The position is estimated by arithmetic calculations of selected matching points (neighbors) in signal space (fingerprint maps). Some improvement is weighted version of kNN—wkNN algorithm. The final position is calculated with use of distances to neighbors scaled with weights reversely proportional to Euclidian distance to that neighbors. The best reported accuracy of these methods is 1 ÷ 3 m (20 % better for kNN, and even better for wkNN). However again, the accuracy depends strongly on environment and measurement conditions.

The works presenting methods that apply statistical learning theory, e.g. [17, 18], report the accuracy similar to wkNN. In the work [17], in tested environment, the 90 ‰ has a distance estimation error equal 5.12 m while the wkNN 5.16 m. Reported errors for neural network based methods varies. In the work [19] neural network classifier to determine location was proposed. Examined average distance error was equal 1.5 ÷ 3 m. In [9] the 90 ‰ has a distance estimation error for this method equal 5.4 m.

3 Weighted kNN-Bayes Algorithm

The probabilistic methods have its basis in the nature of a radio signal. Strong interference, reflections, distortion causes the signal in one place is not homogeneous. Instead, most frequently occurring values of signal strength can be determined. The general idea is that for given set of localization candidates $L = \{l_1, \ldots, l_n, \ldots, l_N\}$ and a measured vector of signal strength $S(s_1, \ldots, s_i, \ldots, s_I)$, we choose the localization l_n for which the probability $P(l_n|S)$ of occurring measured values of S is the biggest:

$$\mathrm{argmax}_{l_n} P(l_n|S) \tag{1}$$

Probabilistic methods use histograms of particular values of radio signal strength, others are based on creating empirically probability distributions and some use ready-made mathematical models of histograms [20, 21].

Our proposed Weighted kNN-Bayes (wkNN-Bayes for short) algorithm is modification of classical probabilistic approach using normal distribution for the estimation of localization. The wkNN-Bayes algorithm combines k nearest neighbor technique with probabilistic algorithm based on Bayes theory using normal distribution to model signal strength distribution.

According to Bayes theory the Eq. (1) can be expressed as:

$$\mathrm{argmax}_{l_n} P(l_n|S) = \mathrm{argmax}_{l_n} \left(\frac{P(S|l_n) \cdot P(l_n)}{P(S)} \right) \tag{2}$$

where $P(l_n)$ is the probability of being in location l_n. Due to the inability to determine its previous location it can be assumed that the probability $P(l_n)$ are equally for all localizations. Also $P(S)$ is constant, so the equation for finding position of the object can be expressed as:

$$\mathrm{argmax}_{l_n} P(l_n|S) = \mathrm{argmax}_{l_n} P(S|l_n) \tag{3}$$

To avoid burdensome determination of the probability of occurrence of a given signal strength for various Wi-Fi networks, as a probability distribution can be assumed normal distribution with empirically determined standard deviation. Then the rule to estimate object localization is:

$$\mathrm{argmax}_{l_n} P(l_n|S) = \mathrm{argmax}_{l_n} \prod_{i=1}^{K} f_{\bar{s}_{i,\sigma}}(s_i) \tag{4}$$

where: $f_{\bar{s}_i,\sigma}$—the normal distribution function with mean distribution $\bar{s}_i$ equals the average value of ith signal strength (i.e. the signal strength of ith access point) at localization l_n, s_i—measured signal strength, and σ—standard deviation of normal distribution.

In our approach we propose weighted version of described Bayes algorithm and using k nearest neighbor technique. The estimated position is calculated as an average of k most likely neighbors $l_{n,k}$, and the probability of given localization in (4) is scaled with the factor (weight) $w_{s|ln}$:

$$w_{S|l_n} = \ln\left(\frac{1}{P(S|l_n)}\right) \tag{5}$$

The weight is logarithm of inverse of the probability. The final position is calculated as an weighted average from k most likely positions. Thus it makes more probable localizations larger impact on calculation of estimated final localization.

4 Experimental Study

One of the main purpose of measurements was to test effectiveness of selected Wi-Fi indoor localization methods with personal localization system in non-controlled active environment. The taken assumptions were the following:

- the testing environment will be a large shopping center with private Wi-Fi networks (with unknown localization of Wi-Fi signal transmitters),
- the measurements will be performed during opening hours when moving people generate additional Wi-Fi signal interference,
- ordinary popular mobile phones be used for measurements will.

The experiment for evaluation of effectiveness of tested algorithms was performed in following steps:

- preparation of detailed map of tested area with geographical data,
- vectorization of the map to use it in measurement tool (mobile phone application),
- performing test measurements of offline phase to build fingerprint map,
- performing online phase measurements to collect data for evaluation of localization,
- data analysis for evaluation of tested algorithms accuracy.

The selected area for experiment was the part of one floor of Grunwald Passage (Pasaz Grunwaldzki in polish) shopping center in Wroclaw, Poland. The map of the center was calibrated in Polish Geographic Coordinate System PUWG 1992 (EPSG:2180) with use of application C-Geo. Next, the map was transformed to the variant of Mercator projection—WSG 84 Pseudo Mercator (EPSG:3857) which is de facto standard for Web mapping applications (e.g. used by Google Maps service). The measurements were performed with use of HTC Desire S mobile phone (smartfon). Its precision of measuring the strength of Wi-Fi signal is 1 dBm, and the measurement frequency is 1 s. The measurements were performed with specially designed measurement system Ilfar—specialized android application cooperating with server application for storing measured data.

The measurement part of the experiment had the following characteristic:

- the measurements were collected in the area of 2600 m^2,
- the offline phase measurements were collected in 143 points presented in Fig. 1,

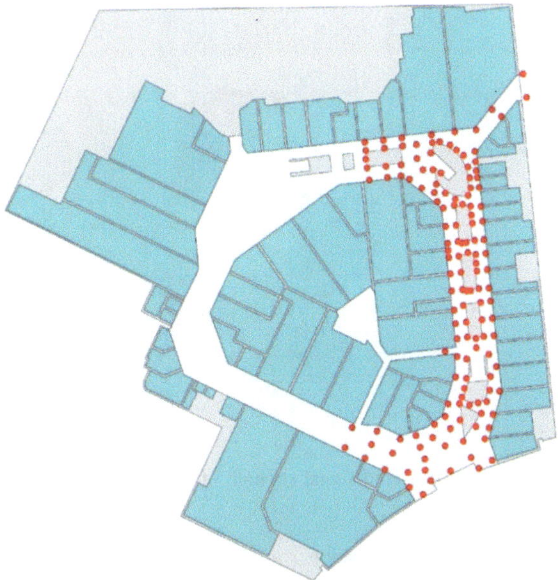

Fig. 1 The map of measurement area (shopping center) with fingerprint measurement points

- each fingerprint (point of map) was calculated from average a number of series of measurements (approx. 8200 for one map) preformed in all directions,
- the average number of access points in one measurement point was 14.65,
- the online measurements (measurements for evaluation of algorithms accuracy) were collected in 116 different points (known precise positions) in the same area,
- the values of signal strength was in the range of −40 to −98 dBm.

The measurements resulted in the maps of the signal strength of various networks—two examples are shown in Fig. 2 (the strongest signal for green (light grey for b&w printing) area, the weakest for dark red/brown (dark grey for b&w printing) area). The collected data during online phase were used to compare accuracy of the following fingerprint localization algorithms: NN—nearest neighbor, kNN—k nearest neighbors, wkNN—weighted k nearest neighbors, Bayes—described in Sect. 3 Bayesian algorithm, wkNN-Bayes—proposed modification of Bayesian algorithm.

At the first glance the maps shows the expected non-monotonic change of signal strength what will most likely affect localization accuracy.

The proposed wkNN-Bayes algorithm was first tested to find the best value of standard deviation of normal distribution σ. The analysis performed for five access points used to determine the localization showed that for σ equal 1 ÷ 1.2 the error of localization stops to decrease. The value of σ = 1.2 was taken for all further experiments. Figure 3 presents, often used in the literature for comparisons, the

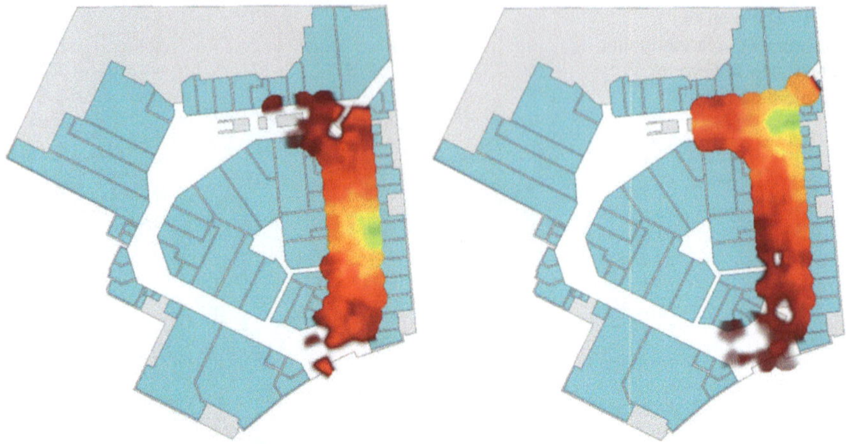

Fig. 2 The measured signal strength of "OMEGA" (*left*) and "orange hotspot" (*right*) networks

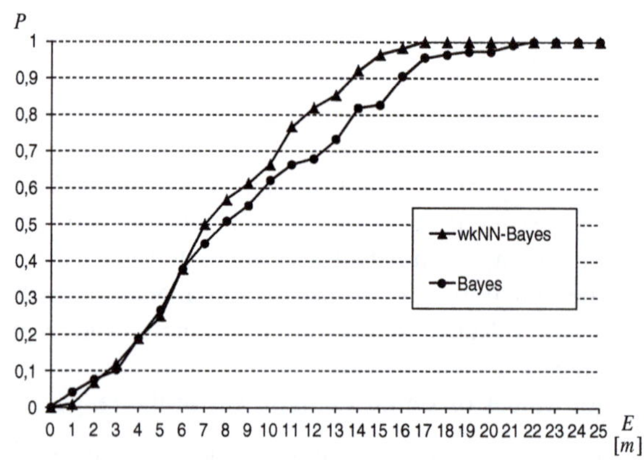

Fig. 3 CDF of the position estimation error E for Bayes and wkNN-Bayes algorithms

empirical distribution function (empirical CDF) of the position estimation error E for Bayes algorithm and proposed wkNN-Bayes algorithm. The experiment showed the apparent advantage of our algorithm over the classic Bayes. The probability P that the estimation of localization error will be less then given value increases especially starting from 6 m. The improvement requires really little additional processing and no extra measurements comparing to Bayes algorithm. The empirical CDF of other kNN algorithms were very close to wkNN-Bayes.

The comparison of all tested algorithms is presented in Fig. 4. The average error of estimation of localization E_{avg} was calculated for different number k of signal transmitters (Wi-Fi access points) i.e. the number of neighbors used to estimate the position of mobile object.

The results show clear advantage of kNN, wkNN and wkNN-Bayes algorithms. However, it did not show the expected differences between known algorithms

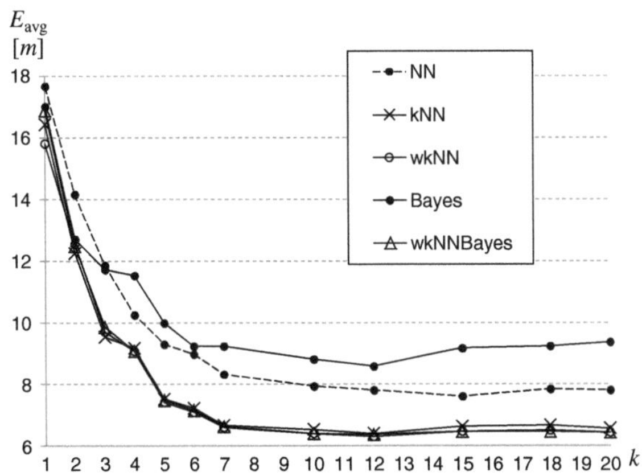

Fig. 4 The average error of estimation E_{avg} versus the number k of neighbors

indicated in the literature. Several possible reasons may explain this fact. The one is that for really hard environment (i.e. when signal interferences are high) the differences between more advanced algorithms are point-sized. The second, that differences in generation of fingerprint map, i.e. map precision and density of measurement points, are so strongly relevant. Also using ordinary measurement can make a difference. The last one is the large number of access points and not using any preprocessing. As shown in [22] for strong signals the measured signal strength variation is greater and any interferences have more significant impact on accuracy. For such a large number of transmitters the mobile object is usually within a range of strong signals.

The proposed wkNN-Bayes algorithm works as well as the best of the rest. At the same time we can see that for most methods in our tested environment the good number of access points used to estimation of the position is 7–12.

5 Final Remarks

The paper focused on two issues: examination of effectiveness of selected indoor localization methods based on Wi-Fi fingerprint maps and presentation of improvement for classic probabilistic algorithm based on Bayes theory. One of the main conclusions from performed experiments is that the impact of not always considered test environment conditions on evaluation of algorithm accuracy is really huge. The overall accuracy of all tested algorithm was not as high as we would expect. Such factors as laboratory vs. live environment and the characteristic of interferences, but also density of fingerprint points and Wi-Fi access points, should be taken into account to compare proposed solutions, because it may impact accuracy of methods distinctly.

The proposed wkNN-Bayes algorithm combines classical approach based on Bayes theory with use of normal distribution to model signal strength distribution and weighted k nearest neighbor approach. It works evidently better than the basal. Averaging estimation of localization improved accuracy, partly overcoming irregularity of Wi-Fi signal. However, without some improvements methods may not work well in active non-controlled environment in the presence of dynamically changing interferences. Especially using preprocessing phase for access point selection seem to be necessary step. Also very careful preparation of fingerprint maps matter.

References

1. U.S. DoD Positioning, Navigation, and Timing Executive Committee: GPS Standard Positioning Service (SPS) Performance Standard, 4th edn (2008)
2. U.S. DoD Positioning, Navigation, and Timing Executive Committee: GPS Precision Positioning Service (PPS) Performance Standard, 1st edn (2007)
3. Monteiro, L.S., Moore, T., Hill, C.: What is the accuracy of DGPS? J. Navig. **58**(2), 207–225 (2005)
4. Otsason, V., Varshavsky, A., LaMarca, A., de Lara, E.: Accurate GSM indoor localization. In: Beigl, M. et al (eds.) UbiComp 2005: Ubiquitous Computing, LNCS, vol. 3660, pp. 141–158. Springer, Heidelberg (2005)
5. Ni, L.M., Liu, Y., Lau, Y.C., Patil, A.P.: LANDMARC: indoor location sensing using active RFID. Wirel. Netw. **10**(6), 701–710 (2004)
6. Tesoriero, R., Gallud, J.A., Lozano, M., Penichet, V.M.R.: Using active and passive RFID technology to support indoor location-aware systems. IEEE Trans. Consum. Electron. **54**(2), 578–583 (2008)
7. Correal, N.S., Kyperountas, S., Shi, Q., Welborn, M.: An ultra wideband relative location system, In: Proceedings of IEEE Conference on Ultra Wideband Systems and Technologies, pp. 394–397, IEEE (2003)
8. Liu, H., Darabi, H., Banerjee, P., Liu, J.: Survey of wireless indoor positioning techniques and systems. IEEE Trans. Syst. Man Cybern. Part C Appl. Rev. **37**(6), 1067–1080 (2007)
9. Torres-Solis, J., Falk, T.H., Chau, T.: A review of indoor localization technologies: towards navigational assistance for topographical disorientation. In: Molina, F.J.V. (ed.) Ambient Intelligence, pp. 51–83. InTech (2010)
10. Mok, E., Retscher, G.: Location determination using WiFi fingerprinting versus WiFi trilateration. J. Locat. Based Serv. **1**(2), 145–159 (2007)
11. Kim, B., Wonsun, B., Kim, Y.C.: Indoor localization for Wi-Fi devices by cross-monitoring AP and weighted triangulation. In: Proceedings of the Consumer Communications and Networking Conference (CCNC) 2011, pp. 933–936, IEEE (2011)
12. Wibowo, S.B., Klepal, M., Pesch, D.: Time of flight ranging using off-the-self IEEE802. 11 WiFi Tags. In: Proceedings of the International Conference on Positioning and Context-Awareness (PoCA 2009), Antwerp, Belgium (2009)
13. Lanzisera, S., Zats, D., Pister, K.S.J.: Radio frequency time-of-flight distance measurement for low-cost wireless sensor localization. IEEE Sens. J. **11**(3), 837–845 (2011)
14. Pahlavan, K., Li, X., Makela, J.P.: Indoor geolocation science and technology. Commun. Mag. **40**(2), 112–118 (2002)
15. Bahl, P., Padmanabhan, V.N.: RADAR: an in-building RF-based user location and tracking system. In: Proceedings of Nineteenth Annual Joint Conference of the IEEE Computer and Communications Societies, INFOCOM 2000, vol. 2, pp. 775–784, IEEE (2000)
16. Bahl, P., Padmanabhan, V.N., Balachandran, A.: Enhancements to the RADAR user location and tracking system, Tech. Rep. MSR-TR-2000–12, Microsoft Corp. (2000)

17. Brunato, M., Battiti, R.: Statistical learning theory for location fingerprinting in wireless LANs. Comput. Netw. **47**(6), 825–845 (2005)
18. Wu, C.L., Fu, L.C., Lian, F.L.: WLAN location determination in e-home via support vector classification, In: Proceedings of the IEEE International Conference on Networking, Sensing and Control 2004, vol. 2, pp. 1026–1031, IEEE (2004)
19. Battiti, R., Villani, A., Le Nhat, T.: Neural network models for intelligent networks: deriving the location from signal patterns. In: Proceedings of the IEEE Symposium on Autonomous Intelligent Networks and Systems AINS2002, UCLA (2002)
20. Roos, T., Myllymaki, P., Tirri, H., Misikangas, P., Sievanen, J.: A probabilistic approach to WLAN user location estimation. Int. J. Wirel. Inf. Netw. **9**(3), 155–164 (2002)
21. Youssef, M.A., Agrawala, A., Shankar, A.U.: WLAN location determination via clustering and probability distributions. In: Proceedings of the First IEEE International Conference on Pervasive Computing and Communications 2003 (PerCom 2003), pp. 143–150, IEEE (2003)
22. Kaemarungsi, K., Krishnamurthy, P.: Modeling of indoor positioning systems based on location fingerprinting, Proc. of 23rd AnnualJoint Conference of the IEEE Computer and Communications Societies (INFOCOM 2004), vol. 2, pp. 1012–1022, IEEE (2004)

The Methodology for Software Development to Support Decision Making Processes

Grzegorz Kołaczek, Paweł Świątek, Krzysztof Juszczyszyn, Adam Grzech and Arkadiusz Sławek

Abstract The paper describes an original software development Methodology and its supporting online tools (the Process Optimization Platform) which allow efficient identification of optimization problems in transport organizations and fast development of a prototype solution. The result of applying the Methodology and the Platform in the context of an organization is a prototype of a decision support system, which can be evaluated with a sample of live data representing actual problems of the organization. This enables the organization to easily assess potential business benefits and quality of the solution before engaging in full software development process related to the implementation of decision support system. The benefits brought by the application of cloud computing solutions while implementing and applying the proposed set of methods and tools have been presented.

Keywords Resource optimization · Decision making · Software development

G. Kołaczek (✉) · P. Świątek · K. Juszczyszyn · A. Grzech · A. Sławek
Chair of Computer Science, Wrocław University of Technology,
Wybrzeże Wyspiańskiego 27, 50-370 Wrocław, Poland
e-mail: grzegorz.kolaczek@pwr.edu.pl

P. Świątek
e-mail: pawel.swiatek@pwr.edu.pl

K. Juszczyszyn
e-mail: krzysztof.juszczyszyn@pwr.edu.pl

A. Grzech
e-mail: adam.grzech@pwr.edu.pl

A. Sławek
e-mail: arkadiusz.slawek@pwr.edu.pl

© Springer International Publishing Switzerland 2016

J. Świątek et al. (eds.), *Information Systems Architecture and Technology: Proceedings of 36th International Conference on Information Systems Architecture and Technology – ISAT 2015 – Part III*,
Advances in Intelligent Systems and Computing 431, DOI 10.1007/978-3-319-28564-1_2

1 Introduction

The Methodology of the Process Optimization Platform focuses on identification, analysis, planning and optimization of enterprise resource management processes in domain-specific information systems. The first part of the paper describes characteristics of basic elements associated with the proposed Methodology. Goals of identification and description of activities in the domain of transport have been described, as well as usage of systematic knowledge in the form of dictionaries and repositories of data gathered from previously conducted studies. The second part of the paper includes details of four classes of use-cases, so-called "Methodology paths" [1, 2].

Main features characterizing the developed Methodology and distinguishing the proposed approach from other solutions are [3]:

- agile approach—involving the provision of methods and tools to make rapid and accurate identification of the relevant decision-making processes for the organization,
- accumulation of knowledge—the knowledge derived from previously solved problems is collected and made available for future systematic analysis of organizations,
- use of service-oriented architecture—use of the SOA paradigm provides ability to streamline the implementation phase and facilitate the integration of newly developed software solutions and organizations' legacy information systems.

2 The Methodology

The Methodology proposes a complete procedure to be followed by development teams. It defines a series of steps, from the very beginning of business analysis, through the structured specification of an optimization problem, to the final delivery of a prototype solution—a software implementation of an algorithm solving the identified optimization problem. The Methodology also defines a set of structures and software elements which are required while following the procedure defined by the Methodology [4–6]. These elements are as follows:

- a tool to assist work of an analyst,
- a method of identification of optimization needs,
- a collection of dictionaries and corresponding development tools,
- a repository of tasks decision models,
- a method of selecting or generating algorithms solving optimization problems,
- a repository of algorithms solving optimization problems.

From its early beginnings the Methodology shares the same principles and ideas which lay behind service oriented and agile computing—sharing resources to achieve coherence and economies of scale, similar to a utility over a network. The convergence between the Methodology and service oriented architecture appears clearly when we consider the common aim which is maximizing effectiveness of shared resources. In case of the Methodology resources are defined by:

- software tools to facilitate business analysis process which can be simultaneously shared by multiple users and applied in different context of organization needs,
- aggregation of domain knowledge gathered during previous analyses,
- aggregation of the algorithms solving various optimization problems identified in organizations,
- service orientation, so that different tools and components supporting problem identification, description and software development process can be exchanged and combined to build new functionalities.

In addition to the advantages of utilizing dedicated online services to support identification and analysis of optimization problems, an important benefit to the Methodology comes from building cloud based computation environments. The ability of diffusion and concurrent computations combined with flexibility of resource management and on-demand availability offered by cloud oriented environments brings a straightforward improvement to the sophisticated computation methods required by models and algorithms forming the final solutions of identified optimization problems.

3 Main Elements Defining the Methodology

The aim of processes defined by the Methodology (Fig. 1) in addition to identifying organization's optimization needs is to prepare a prototype solution. Depending on analyzed context, preparation of a prototype requires selection of one of four possible paths defined by the proposed Methodology.

3.1 Finding Solutions

The first path defined by the Methodology—"Finding Solutions"—is followed if an adequate solution of decision making task already exists and is available in one of the Process Optimization Platform repositories. It means that for a given organization and its identified optimization problem, a solution algorithm can be found in

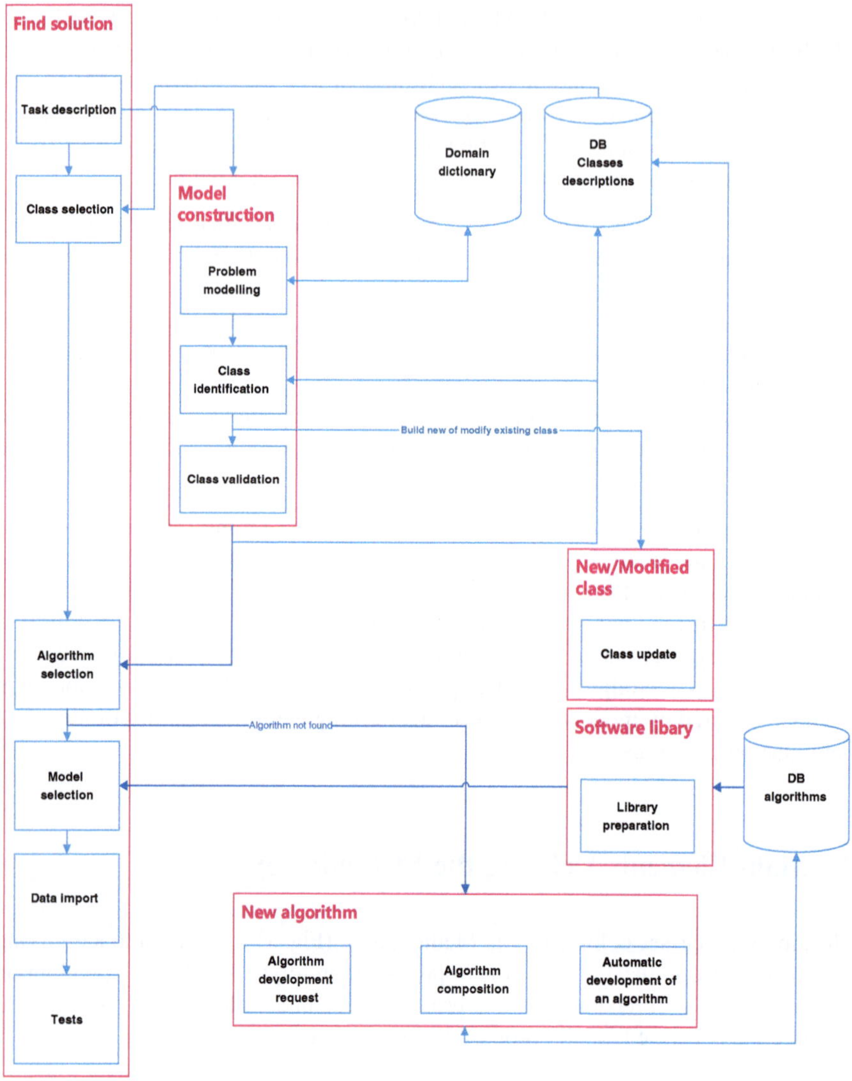

Fig. 1 The methodology

the Platform repository and matched based on the similarity of the current problem to one of previously analyzed cases. In this scenario, we assume that the matched algorithm can be re-used in the preparation process of prototype solution for the given organization. In this path of the Methodology all decisions regarding the choice of the model and the algorithm class are undertaken independently by an appropriate system analyst (Fig. 2).

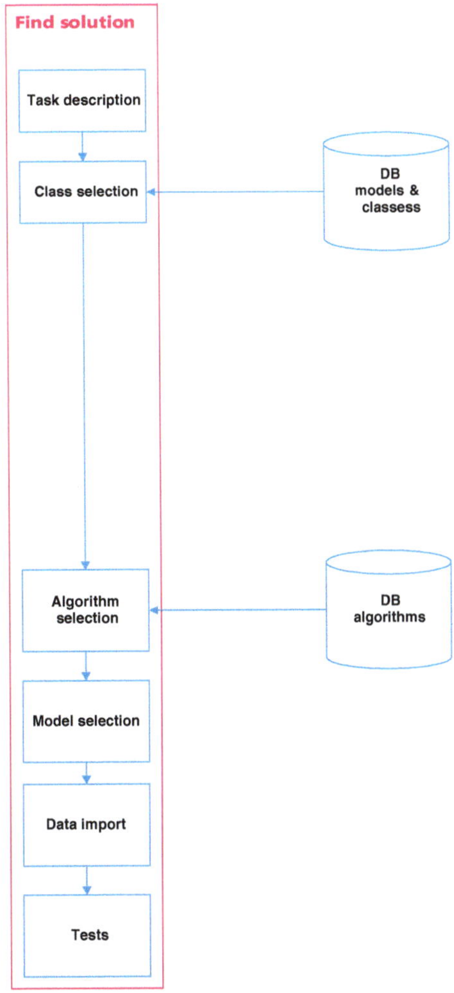

Fig. 2 Path 1: finding solutions

3.2 *Path 2: Model Construction*

The second path in the Methodology—"Model construction"—is followed when an analyst builds a description of an organization using online Platform modeling tools (Fig. 3). The Process Optimization Platform comprises online tools dedicated to support the structured process of gathering and organizing knowledge about an organization, its decision making processes and occurring optimization problems. The Platform, supplied with a set of tools, allows for interactive use of domain knowledge accumulated in its repositories and domain dictionaries. The goal of organization modeling process is to enable automation of task model definition and aided selection of relevant classes for the defined optimization tasks. As the

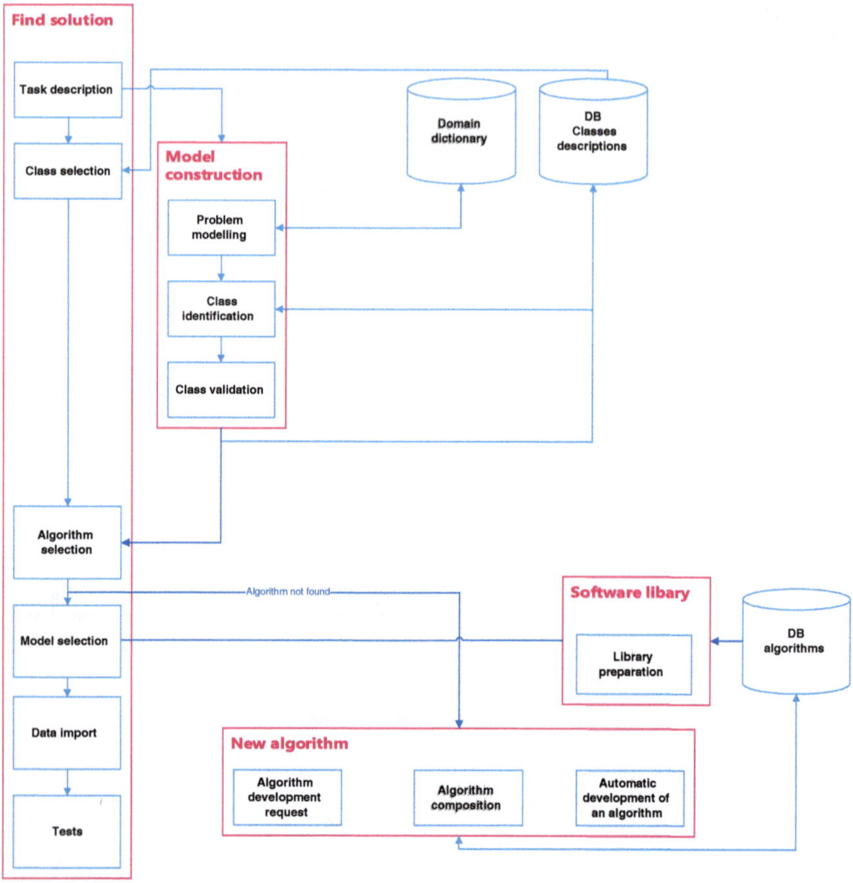

Fig. 3 Path 2: model construction

Platform repositories maintain relations between gathered solution algorithms and optimization task classes, selecting a proper class enables instant discovery of matching solution algorithms and further—generation of a prototype. Path 2 describes scenarios in which, like in the case of Path 1, there are appropriate models, class definitions and algorithms already available in the repositories of the online Platform.

3.3 Path 3: Modification of a Class

The third path in the Methodology—"Modification of a class"—is selected if an analyst using the modeling tools provided by the online Platform states that available types and classes of tasks are not sufficient to properly reflect the optimization problem of the analyzed organization and it is not possible to precisely define the optimization task (Fig. 4). The lack of a class which would directly define the analyzed optimization process occurs when the optimization problem encountered differs from any of previously analyzed cases but some general similarities can be found. Although there is no class definitions in the Platform repository ready to be used directly, classes related to the same domain do exist. Therefore, in the Path 3 of the Methodology, there is a need to update the definition of an available class to match the analyzed optimization problem and a need to build new solution algorithm. According to the general concept of the Platform, a new solution algorithm can be made available as a result of one of three activities. The first possible scenario is when a new algorithm is implemented by a specialist and uploaded to the Platform repository. It is important, that while a specialist is preparing an implementation of a solution algorithm he/she uses the detailed guidance provided by the Platform and the model of the organization built with the

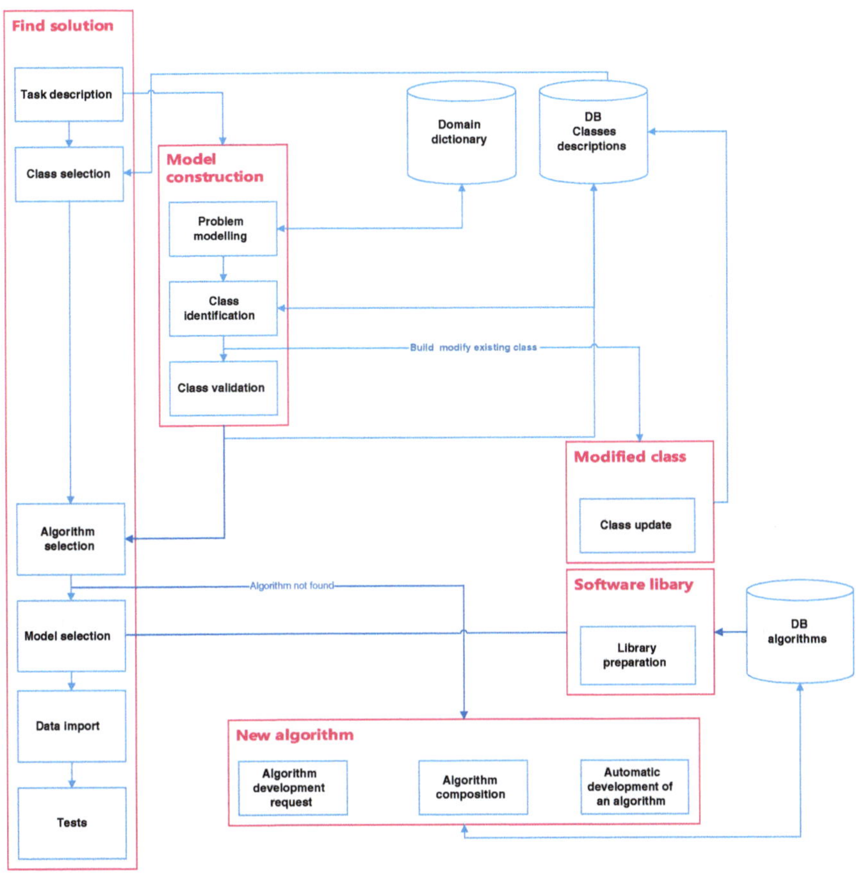

Fig. 4 Path 3: modification of a class

Platform tools. The second method to deliver a solution algorithm is related to the so-called manual composition algorithm. It means that a software engineer configures and publishes a composite service which fulfills all requirements defined during earlier stages of analysis. In the third variant a solution algorithm is composed and provided automatically by mechanisms implemented in the Platform.

3.4 Path 4: Defining a New Class

The fourth path of the Methodology—"Defining a new algorithm"—describes a class of scenarios in which, like it was in the case of the Path 3, the analysis identified a new, previously not described problem of organization's optimization needs (Fig. 5). In this case however, unlike in the Path 3, there are no classes available or

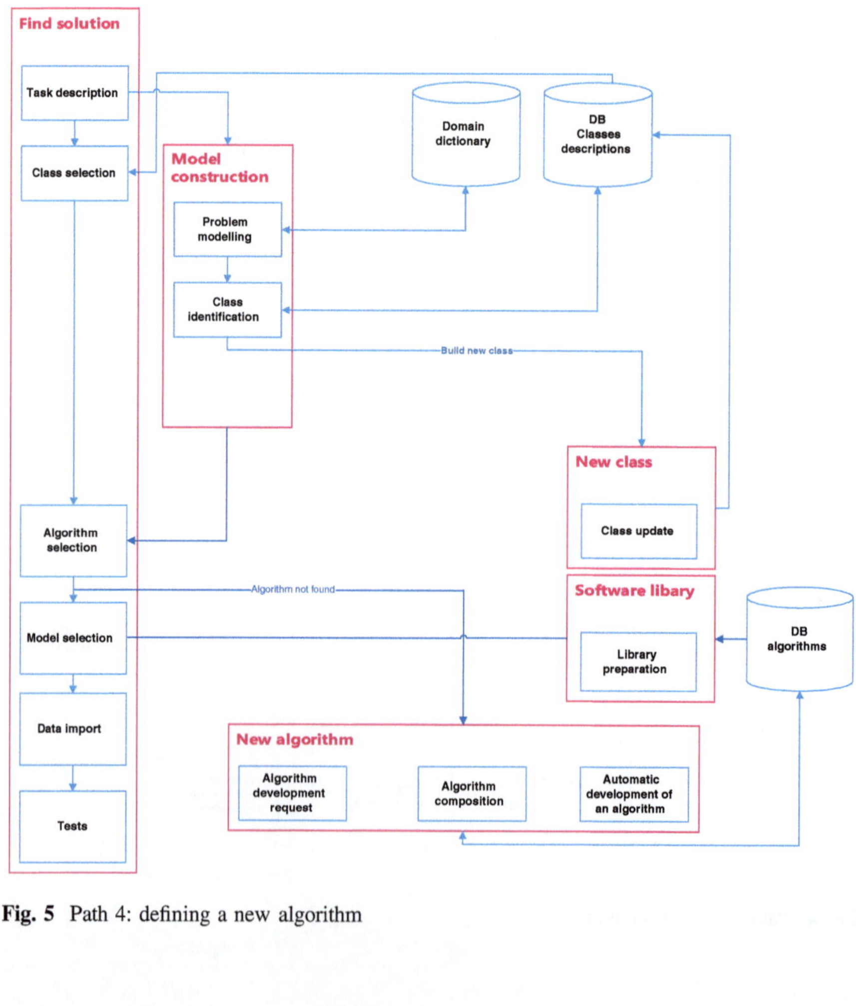

Fig. 5 Path 4: defining a new algorithm

none of the classes can be modified to match the identified problem. Therefore, the primary action that must be taken in the context of this path, is to add a new class of optimization problems to the Platform repository, matching the definition of the analyzed task. Also, in addition to extending the task class repository, there is a need to provide a new algorithm solving the new optimization problem. The algorithm is made available on a similar basis as described in the case of Path 3, i.e. the algorithm can be prepared by an expert, can be made available as a result of the so-called manual composition, or also in certain cases it is possible to automatically compose an algorithm using the techniques and methods available with the Platform.

4 Conclusions

The Methodology proposes a complete procedure to be followed by development teams. It defines a series of steps, from the very beginning of business analysis, through the structured specification of an optimization problem, to the final delivery of a prototype solution—a software implementation of an algorithm solving the identified optimization problem. To the best authors' knowledge the Methodology and the Platform are the first available solution which delivers holistic support for specialized software development for supporting decision making processes in business organizations. The Methodology also defines a set of structures and software elements which are required while following the procedure defined by the Methodology.

The key benefits offered by using the proposed Methodology in the production of software for the analysis, planning and optimization of resource management processes of the organization are:

- quicker development, implementation and adaptation of the system by the possibility of using libraries of standard and specialized processes and algorithms dedicated to the areas of transport management,
- reduction in the system implementation costs by simplifying and speeding up the analysis of the requirements for the development or adaptation of the system,
- optimization of organization's management processes and utilization of resources by selecting appropriate optimization methods dedicated to specific customer requirements,
- integration of optimization modules and functionality of monolithic and legacy systems through the use of service-oriented paradigm (Service Oriented Architecture).

The Methodology has been developed and evaluated in the context of the optimization needs of transportation organizations. However, it is possible to benefit from the Methodology and the Platform also in many other areas of business activities. The future works include e.g. application of developed tools and methods to support production process and warehouse management in small and medium-sized enterprises.

Acknowledgment This work has been partially supported by the Polish Ministry of Science and Higher Education within the European Regional Development Fund, Grant No. POIG.01.03.01-02-079/12.

References

1. Brzostowski, K., Gąsior, D., Grzech, A., Juszczyszyn, K., Kołaczek, G., Kozik, A., Rudek, R., Sławek, A., Szczurowski, L., Świątek, P.R.: Business process optimization platform for integrated information systems, information systems architecture and technology. In: Świątek, J. (ed.) [i in.], pp. 13–22. Oficyna Wydawnicza Politechniki Wrocławskiej, Wrocław (2014)
2. Kołaczek, G., Juszczyszyn, K., Świątek, P., Grzech, A.: Decision making in security level evaluation process of service-based applications in future internet architecture. In: Advances in Systems Science, pp. 701–714. Springer International Publishing, Berlin (2014)
3. Grzech, A., Juszczyszyn, K., Świątek, P.: Methodology and platform for business process optimization. In: Advances in Intelligent Systems and Computing—Progress in Systems Engineering, pp. 377–382. Springer, Berlin (2015)
4. Juszczyszyn, K., Kołaczek, G., Świątek, P., Grzech, A.: Agile method of software development for resources management optimization processes. In: Information Systems Architecture and Technology: Systems Analysis Approach to the Design, Control and Decision Support, pp. 53–62. Oficyna Wydawnicza Politechniki Wrocławskiej, Wrocław (2014)
5. Jinghai, R., Xiaomeng, S.: A survey of automated web service composition methods, semantic web services and web process composition, pp. 43–54. First International Workshop, SWSWPC, San Diego (2004)
6. Ko, J.M., Kim, C.O., Kwon, I.K.: Quality-of-service oriented web service composition algorithm and planning architecture. J. Syst. Soft. **81**, 2079–2090 (2008)

Security Lock System for Mobile Devices Based on Fingerprint Recognition Algorithm

Michał Szczepanik, Ireneusz J. Jóźwiak, Tomasz Jamka and Karol Stasiński

Abstract In this paper authors focus on fingerprint recognition biometric systems used by mobile devices to unlock or to confirm user actions. Authors compare the existing fingerprint recognition algorithms and test their own algorithm on fingerprints database, which include changes in the structure as a result of physical damage. They also propose new selective attention algorithms, which help to detect the most sensitive to damage areas, and add it as step of fingerprint analyses for the fingerprint recognition procedures. They also propose a new algorithm, which does not require complex hardware systems, so it can be applied in mobile devices, which restrict unauthorized access to sensitive data or other user's resources. The main goal of this work is to demonstrate the applicability of the developed algorithm for mobile devices.

Keywords Biometric · Fingerprint · Minutia group · Access controls

1 Introduction

Nowadays, mobile phones, tablet PCs and other mobile devices are an excellent source of data about users. They not only store information about user contacts, but also history of user's live, user's habits, detail information about user's friends and family. They are used as a remote office, email client, for bank account management

M. Szczepanik (✉) · I.J. Jóźwiak · T. Jamka · K. Stasiński
Faculty of Computer Science and Management, Department of Informatics, Wrocław University of Technology, Wrocław, Poland
e-mail: michal.szczepanik@pwr.edu.pl

I.J. Jóźwiak
e-mail: ireneusz.jozwiak@pwr.edu.pl

T. Jamka
e-mail: tomek.jamka@gmail.com

K. Stasiński
e-mail: karol.stasinski@gmail.com

© Springer International Publishing Switzerland 2016

J. Świątek et al. (eds.), *Information Systems Architecture and Technology: Proceedings of 36th International Conference on Information Systems Architecture and Technology – ISAT 2015 – Part III*, Advances in Intelligent Systems and Computing 431, DOI 10.1007/978-3-319-28564-1_3

as well as for entertainment like social media and games. Typically applicable security method is a four-digit pin or graphical pattern, which is usually easy to guess or break and block only access to mobile network. Computational capabilities of current devices are not as limited as in 2008–2010 when most of current systems were designed [1].

Today, mobile phones have the computational capabilities similar to personal computers which could be bought one or two years ago; so, why should they not use better security systems, such as biometrics systems. The accuracy of a fingerprint verification system is critical in a wide variety of civilian, forensic and commercial applications such as access control systems, credit cards, phone usage, etc. The main problem, from an economic point of view, can be the size of the used system hardware and its cost, and therefore it cannot be too extensive and advanced.

Fingerprints are the most widely used biometric feature for personal identification and verification in the field of biometric identification [2, 3]. Most important for designing the system is the effectiveness of fingerprint recognition algorithms, which depends mainly on the quality of digital fingerprint images input and fingerprint's physical damage [4].

Current mobile devices typically use as collateral for a four digit pin code or face recognition system, which doesn't work when is too darker. Most fingerprint recognition algorithms are not immune to damage, so they are not use in mobile devices. The main problem, which we would like to solve is the stability of the recognition systems with respect to the capability to deal with fingers' damages, will make the system more useful to the user.

2 Quality Assessment of Biometric Algorithms

There are two most important performance metrics for biometric systems [2, 5]: FAR and FRR.

False Accept Rate (FAR), also called False Match Rate (FMR), is the probability that the system incorrectly matches the input pattern to a non-matching template from the database. It measures the percent of invalid inputs which are incorrectly accepted.

False Reject Rate (FRR), also called False Non-Match Rate (FNMR), is the probability that the system fails to detect a match between the input pattern and a matching template from the database. It measures the percent of valid inputs which are incorrectly rejected. They can be presented mathematically as:

$$\mathrm{FAR(T)} = \int_{\mathrm{Th}}^{1} \mathrm{g(x)dx} \tag{1}$$

$$\mathrm{FAR(T)} = \int_{0}^{\mathrm{Th}} \mathrm{g(x)dx} \tag{2}$$

where Th is the value of a threshold used in the algorithm and g(x) is function of algorithm that analyzed all samples with pattern. Both FAR and FRR are functions of a threshold T. When T decreases, the system has more tolerance to intraclass variations and noise, however, FAR increases. Similarly, if it is lower, the system is more secure and FRR decreases.

3 History and Existing Solutions

First mobile phone with fingerprint recognition system was developed by Siemens in 1998 [3]. Since that time, more than 100 phone models had such a protection [5] and this type of system start to be more popular. Unfortunately, the biggest problem of those systems was usability.

Every day, people are exposed to cuts, wounds and burns; therefore, it is important that the algorithms are resistant to this type of damage. Existing fingerprint recognition systems for mobile devices usually use one of the algorithms:

- Minutiae Adjacency Graph (MAG),
- Elastic Minutiae Matching (EMM),
- Delaunay Triangulation (DT),
- Pattern-Based Templates (PBT).

The most popular algorithms are based on local and global structures represented by graphs like in MAG. In this type of algorithms, local structures to find corresponding points to align feature vector are used first, then global structures are matched [6, 7]. This type of algorithm was used by He and Ou [8], Ross et al. [9]. They also use thin-plate spline (TPS) model to build an average deformation model from multiple impressions of the same finger. Owing to iteratively aligning minutiae between input and template impressions, a risk of forcing an alignment between impressions originating from two different fingers arises and leads to a higher false accept rate. Typically, a minutia matching has two steps:

- registration aligns fingerprints, which could be matched
- evaluation calculates matching scores using a tolerance box between every possibly matched point (minutiae) pairs.

The EMM algorithm typically uses only global matching where each point (minutia) which has a type, like end point or bifurcation, needs to be matched to a related point in the second fingerprint image. Based on elastic deformations which are used to tolerate minutiae pairs that are further apart because of plastic disrotations, and therefore to decrease the False Rejection Rate, so in most popular algorithms authors increase the size of bounding boxes [10] to reduce this problem, but as side effect they got higher False Acceptation Rate (FAR). In this type of algorithms, for elastic match, TSP [11] also can be used, which provides better performance than only one parameter of deformation.

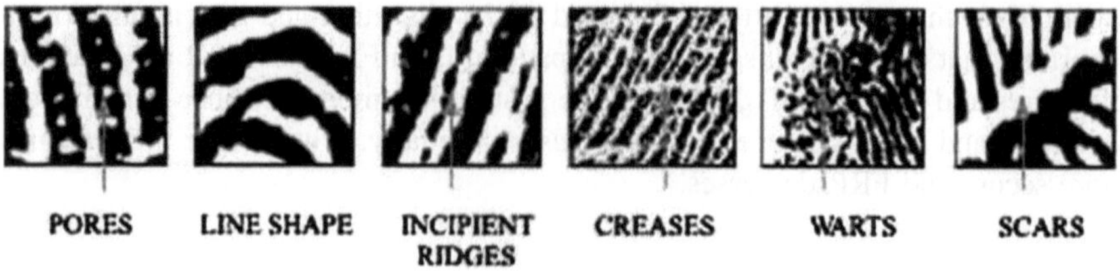

Fig. 1 Typical damages on fingerprint (*source* [4])

The DT algorithm [12, 13] is the most popular version of MAG, so it is described as separate algorithm. Its structure based on triangulation connects neighboring minutiae to create triangles, such that no point (minutia) in P is inside the circumcircle of any triangle in $DT(P)$. DT algorithm analyzes the structure of points identically as minutiae adjacency graph algorithm, so it also is not resistant to typical injury of physical fingerprint [4] (see Fig. 1).

The Pattern based algorithms [14] compare the basic fingerprint patterns (like arch, whorl, and loop) between a previously stored template and a candidate fingerprint. This algorithm requires that the images be aligned in the same orientation and in the same scale. To do this, the algorithm finds a central point in the fingerprint image and centers on it, and after that, scales to the same size of the fingerprint's ridge. In a pattern-based algorithm, the template contains the type, size and orientation of patterns within the aligned fingerprint image. The candidate fingerprint image is graphically compared with the template to determine the degree to which they match. Due to the storage of the original picture for the algorithm there is a high risk that this image can be read from the memory card reader or fingerprints database.

4 Current Solutions

The most popular devices with fingerprints recognition system are iPhone 5 s, iPhone 6, iPhone 6 Plus, iPad Air 2, iPad Mini 3. With solution known as Touch ID, and Samsung S5, Samsung Galaxy Note 4.

Touch ID uses 88-by-88-pixel, 500-ppi raster scan from fingerprint reader which is a part of home button (see Fig. 2). The analysis utilizes sub dermal ridge flow angle mapping, which is a lossy process that discards minutia data that would be required to reconstruct the user's actual fingerprint. This also allow to make this reader more useful in case of different fingerprints orientation.

Samsung's solution based on minutias' structures and hybrid of MAG and EMM algorithm. The reader with this algorithm doesn't allow to work with different fingerprint orientation, so most of users stop use it. The security level is also low as Samsung block this method of authentication when user set any password restrictions in MDM (mobile device management) settings.

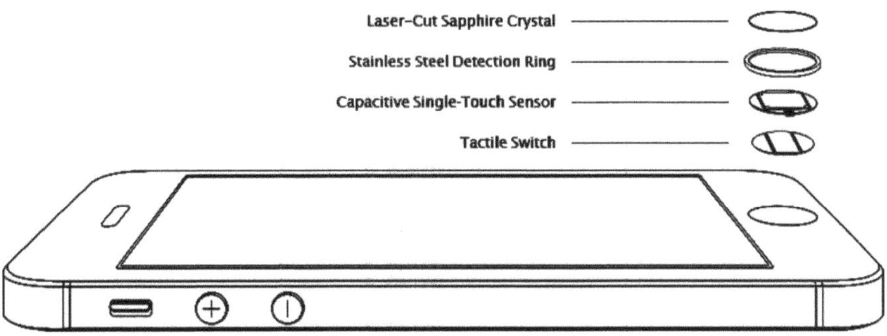

Fig. 2 Touch ID structure (*source* Apple)

Next generation of Android devices should also include fingerprint recognition system for authentications as this type of method is included in Android M preview. There is system's native support for this method and mobile devices manufacturers have open door for them.

5 Fingerprint Recognition Algorithm Based on Minutia' Groups

The proposed solutions, in contrast to other algorithms, are more resistant to damage and it based on minutias' groups instead on single minuta point.

5.1 Preprocessing

Preprocessing step was explained in details in previous publications [15] and whole process is graphically presented on Fig. 3.

For each scanned image is required to detect the areas of correctly scanned parts of the fingerprint. First step of image analysis is the search for the imprint area including the exclusion of areas containing significant damage. Fingerprint image is represented by a gray scale image that defines the area of forced application fingerprint for the reader and can be represent as values from 0 to 255 as $I_{fp}(i, j)$ where i and j are coordinates of point.

$$I_{fp}(i,j) = \langle 0,\ 255 \rangle \tag{3}$$

The operation that converts a grayscale image into a binary image is known as binarization. The binarization process based on adaptive threshold process. Each pixel is assigned a new value (1 or 0) according to intensity mean in a local area and the parameter t_g which excludes poorly read fingerprint areas from the analysis, see Eq. 4.

Fig. 3 Steps of preprocessing. From *left* original image, binarized image, mask (*source* own work)

$$B_{fp}(i,j) = \begin{cases} 0 \text{ for } I_{fp}(i,j) > t_g \\ 1 \text{ for } I_{fp}(i,j) \le t_g \end{cases} \quad (4)$$

The last step is creating the fingerprint mask based on the binarized image. The Mask for the area of a square or rectangle (*X*, *Y*), where *X*, *Y* represents range of vertical and horizontal coordinates. The size is 2.5 wide edges. Mask is also determined by two parameters p_{lo}, which is a limitation that excludes areas with an insufficient number of pixels describing the image, and p_{hi} excludes blurred areas, such as moist. Mask can be represent by Eq. 5.

$$M_{fp}(X,Y) = \begin{cases} I_{fp}(X,Y) \text{ for } p_{lo} \le F_{img}(X,Y) \le p_{hi} \\ 0 \text{ otherwise} \end{cases} \quad (5)$$

where F_{img} is represented by Eq. 6:

$$F_{img}(X,Y) = \sum_{i \in X} \sum_{j \in Y} B_{fp}(i,j) \quad (6)$$

Created mask is used for finding the most damaged area in the fingerprint image.

5.2 Detecting Features and Leveling of the Damage in Segmentations

Standard leveling of damage is carried out by calculating the variance of points and the analysis of brightness. Based on these two parameters, the frequency of furrows is calculated, which is used for each fingerprint image. After applying Gabor filter to highlight the pits and valleys, it uses segmentation in accordance with its size, 2.5 width of segment furrow, the image is redrawn.

After that process fingerprints are continuous and lint. In contrast to the literature, the algorithm does not require additional transformations to find the minutiae, such as converting the width of 1 px furrows. It does not require information about the orientation of minutiae; it only requires the data about its position. Therefore, the resulting image is used to find the edge—the minutiae are located at the intersection of the edge of the furrows. The problem of fingerprint recognition is a complex process, even in laboratory conditions; therefore, if used as a system to control access to the mobile devices, it should be insensitive to certain natural changes or damages in physical structure of fingerprints, which can include: incomplete fingerprint, fingerprint parts which can be injured or burned, blurred, partly unreadable or rotation. In order to detect the most sensitive to damage areas, we use neural network with selective attention technique [16]. This type of neural network is more like an analysis done by a human. This allows us to create a mask of areas vulnerable to damage.

We created 15 different masks, broken down by the type of fingerprint's core also known as fingerprint patterns (arch, whorl and loop) and the type of finger (thumb, index finger, middle finger, ring finger, small finger). Basing on this mask we created a filter, which we use to compare fingerprints where specific minutiae are weighted in the decision process and their score is based on the location on the fingerprints.

5.3 Comparison of Fingerprint and Pattern

Minutiae image is divided into segments, each segment is corresponding minutiae's group is described by parameters (*x*, *y*, *nom*), where *x* and *y* are the coordinates, and *nom* determines the number of minutiae in the group. Additionally, one implementation uses an additional parameter specifying the probabilities of damage in a given segment, which is estimated by a neural network, based on the distribution of areas rejected by the mask described by the formula. Current algorithm implementation searches small groups of minutiae that contain up to 5 minutiae (see Fig. 4). Then, based on the neighboring groups (max 4) creates a new large group. For each, the orientation parameters and the number of characteristic points are recalculated. The last step is to create a matrix of Euclidean distances between the largest groups.

When comparing the use of two parameters: *dx*—the distance defining the difference between groups in the pattern and tested fingerprint, *px*—the threshold of damage occurrence probability (determined by whether the group is under consideration in the analysis), we decide which groups should be compared and we set the priority for them. After that we do the comparison of the groups, which are divided according to the priority, that is defined by the number of minutiae in the group and selective attention (SA) algorithms, which are based on probabilities of

Fig. 4 Fingerprint divided into segments (*source* own work)

damage in a group segment. This provides quick verification of whether the analyzed fingerprint is consistent with the pattern.

6 The Quality of the Algorithm

First test was done using FVC2004 [17] fingerprint databases. For each of four databases, a total of 120 fingers and 12 impressions per finger (1440 impressions) were gathered. Unfortunately, most of the publicly available databases of fingerprints do not include the problem of physical damage, so additionally small damage such as cuts and burns has been generated on each sample. In most cases artificially applied damages cover 5–20 % of the fingerprint. For 10 % of the samples they cover approximately 50 % of the area to simulate severe damage. Table 1 presents

Table 1 The result of experiment using FVC2004 database

Algorithm	FRR (%)	FAR (%)
MAG	0.82	0.65
EMM	1.23	1.15
PBTA	1.73	0.15
MGM	1.32	0.86
MGM SA	0.28	0.16

results of the tests for FVC2004 database for existing algorithms MAG, EMM, PBTA and our algorithm Minutia Group Matching (MGM) and algorithm with selective attention phase (MGM SA).

7 Classification and Data Management Used in the Algorithm

The developed algorithm is based on minutiae groups, where each group is basically represented by the coordinates—x, y and the number of minutiae—*nom* contained in the group. Group covers an area equal to 2.5 the width of the furrow and its coordinates are in the middle of the square which is boundaring this area. Number of minutiae in the group describes its priority. Additionally, a stored parameter defines the probability of damage—p_d in the area represented by the group. In conclusion the group is defined as follows:

$$M_{\{group\}}\colon \{x, y, nom, p_d\} \tag{7}$$

Based on these data, a matrix of Euclidean distances between the groups is created. Data on the characteristic point is limited to its weight (*nom*) and the probability of damage $\mathrm{p_d}$. Finally we obtain:

$$M_{\{group\}}(I)\colon \left\{nom_I, (p_d)_I\right\} \tag{8}$$

$$M_{\{group\}}(I, J)\colon dist\left(M_{\{group\}}(I), M_{\{group\}}(J)\right) \tag{9}$$

where $dist\left(M_{\{group\}}(I), M_{\{group\}}(J)\right)$ is Euclidean distances between the group I and J.

Data stored for analysis to prevent reproduction of the original fingerprint image. Additional storage parameters to estimate the damage allow us to better match fingerprints in the event of damage.

8 The Experiment

The tests were conducted on two devices: Samsung Galaxy S3 (see Table 2) with additional scanner and Samsung S5 (see Table 3). For test authors used real fingerprints. Due to the nature of work, we use real fingerprints. Each user was exposed to frequent damage of fingerprints, like cuts and burns presented on Fig. 1.

All algorithms were compared using 112 different fingerprints and each had 10 samples. Each fingerprint was scanned once and used by specific algorithm separately, so each algorithm used same input.

Table 2 The result of experiment on Samsung Galaxy S3 with additional fingerprint reader

Algorithm	FRR (%)	FAR (%)	Avg. decision time (ms)
MAG	0.80	0.63	132
EMM	1.20	1.15	123
PBTA	1.70	0.25	127
MGM	1.70	0.86	111
MGM SA	0.28	0.16	120

Table 3 The result of experiment on Samsung S5

Algorithm	FRR (%)	FAR (%)	Avg. decision time (ms)
MAG	0.86	0.86	98
EMM	1.40	1.15	100
PBTA	2.10	0.45	100
OWN (embedded)	8.10	0.80	–
MGM	1.80	0.86	95
MGM SA	0.63	0.25	98

Unfortunately existing solution used by Samsung in S5 (OWN) was not able to recognize 8.1 % of tested images, but it also not use preprocessing which orient scanned image as for other tested algorithms.

9 Conclusion

The problem of biometric authentication on mobile devices has been successfully approached. The main problem was captured of fingerprint image and preprocessing of it in way that allow detect abnormal structures and damages. The biggest problem of all solutions used in mobile devices is acceptance of manually prepared fingerprints. In most of test printed images was used and were recognized and accepted by devices as real fingerprints. The plan for future work is to create algorithm which works with sensors to recognize humanity of scanned object. Next generation of android devices with Android M should also include fingerprints recognition system for user authentication and this fact is also promising as this type of systems will be more popular.

References

1. Ross, A., Nandakumar, K., Jain, A.K.: Handbook of multibiometrics (International Series on Biometrics). Springer, Berlin (2011)
2. Jain, A.K., Ross, A., Nandakumar, K.: Introducing to biometrics. Springer, Berlin (2011)

3. Wayman, J.L., Jain, A.K., Maltoni, D., Maio, D.: Biometric systems. Technology, design and performance evaluation, 1st edn. Springer, Berlin (2005)
4. Hicklin, A., Watson, C., Ulery, B.: How many people have fingerprints that are hard to match, NIST interagency report 7271 (2005)
5. Ratha, N.K., Govindaraju, V.: Advances in biometrics: Sensors, algorithms and systems. Springer, Berlin (2007)
6. Chikkerur, S., Govindaraju, V., Cartwright, E.: K-plet and coupled BFS: A graph based fingerprint representation and matching algorithm. LNCS, pp. 309–315 (2006)
7. Grzeszyk, C.: Forensic fingerprint examination marks. Wydawnictwo Centrum Szkolenia Policji, Legionowo (1992). (in Polish)
8. He, Y., Ou, Z.: Fingerprint matching algorithm based on local minutiae adjacency graph. J. Harbin Inst. Technol. **10**(05), 95–103 (2005)
9. Ross, A., Dass, S.C., Jain, A.K.: A deformable model for fingerprint matching. Pattern Recogn. **38**(1), 95–103 (2005)
10. Pankanti, S., Prabhakar, S., Jain, A.K.: On the individuality of fingerprints. In: Proceedings of Computer Vision and Pattern Recognition (CVPR) (2001)
11. Bazen, A.M., Gerez, S.H.: Fingerprint matching by thinplate spline modelling of elastic defromations. Pattern Recogn. **35**, 1859–1867 (2003)
12. Bebis, G., Deaconu, T., Georgiopoulos, M.: Fingerprint identification using delaunay triangulation. In: IEEE ICIIS, pp. 452–459 (1999)
13. Parziale, G., Niel, A.: A fingerprint matching using minutiae triangulation. In: Proceedings of ICBA (2004)
14. Cappelli, R., Lumini, A., Maio, D., Maltoni, D.: Fingerprint classification by directional image partitioning. IEEE Trans. Pattern Anal. Mach. Intell. **21**(5), 402–421 (1999)
15. Szczepanik, M., Szewczyk, R.: Fingerprint identification algorithm (in Polish). KNS **1**, 131–136 (2008)
16. Indovina, M., Uludag, U., Snelick, R., Mink, A., Jain, A.: Multimodal biometric authentication methods: A COTS approach. In: Proceedongs of MMUA (2003)
17. Maltoni, D., Maio, D., Jain, A.K., Prabhakar, S.: Handbook of fingerprint recognition, 2nd edn. Springer, Berlin (2009)

Dynamic Resource Management Middleware for Modern Service Oriented Systems

Łukasz Falas and Krzysztof Juszczyszyn

Abstract Service-oriented architecture is a major architectural pattern used for the development of large scale IT systems. Along with the increase in adaptation of SLAs (Service Level Agreements) in contracts between clients and service providers, these changes lead to new challenges in the field of Quality of Service (QoS) management. Due to loose coupling of web services and the dynamic, non-deterministic nature of such systems, there is a necessity to develop new methods and infrastructure solutions for proper resource management in such systems. This paper presents a proposition of a middleware for quality of service and resource management, which addresses unique challenges related to service-oriented systems that provide functionality through composite services. The proposition contains the description of the architecture of the middleware system, description of request processing workflow and a definition of quality of service and resource allocation problem for introduced system, followed by the proposition of optimization model for this problem.

Keywords Resource allocation · Qos-Aware service composition · Service oriented architecture

1 Introduction

The increase in number of software users and noticeable migration to web-based solutions requires approaches which support scalability, reliability and quality assurance dedicated for these solutions. Nowadays Service-oriented architecture (SOA) is one of the most acclaimed architectural styles, used for building large

Ł. Falas (✉) · K. Juszczyszyn
Department of Computer Science, Wrocław University of Technology,
Wybrzeże Wyspiańskiego 27, 50-370 Wrocław, Poland
e-mail: lukasz.falas@pwr.edu.pl

K. Juszczyszyn
e-mail: krzysztof.juszczyszyn@pwr.edu.pl

© Springer International Publishing Switzerland 2016

J. Świątek et al. (eds.), *Information Systems Architecture and Technology: Proceedings of 36th International Conference on Information Systems Architecture and Technology – ISAT 2015 – Part III*,
Advances in Intelligent Systems and Computing 431, DOI 10.1007/978-3-319-28564-1_4

scale web-based distributed systems and it may delivered the desired features to web-based distributed systems. One of the main goals of service-oriented architecture is to clearly distinct service clients and service providers, who often may represent different organizations operating in different parts of the world. The interaction between service client and service provider is often regulated by contracts which describes functionalities and quality of offered service.

The main component of the system in service-oriented architecture is the Web Service, which can be defined as an autonomous software component delivering a certain functionality that is accessible through a well-defined interface on the Internet. SOA design principles state, that a web service (further referenced to as atomic service) should offer some basic, atomic, functionality and should not perform any complex tasks or invoke other web services. Due to this fact, complex functionalities should be delivered by composite services collaborating in a workflow. Such collaboration is possible due to standardized communication interfaces which enable web service interoperability and is often referenced as a composite service. The process of building a composite service according to user requirements is known as web service composition.

Composability of web services is an important feature which enables software developers and architects to easily design, maintain and even redesign composite services. For instance, a developer may easily redesign the workflow of a composite service by adding or removing web services, because this change is made at workflow level and not in the code implementation. Also, as long as the communication interfaces are maintained, even the implementation of a web service can be changed (for example, a developer may implement a more efficient algorithm inside the web service) without the need for making any changes in the composite service workflow.

In majority of currently developed systems such scenarios for designing and redesigning a composite service can be done by a developer, however one of the development goals for service-oriented systems is the implementation of automated methods, which will analyze user's requirements and compose a service, which matches the defined user requirements. It is worth noting that along with the possibility to design a composite service according to the functional requirements, composite services can be designed according to user's non-functional requirements. It is possible because web services can have many versions which offer the same functionality, but differ in the values of non-functional parameters. This diversification may be achieved by implementing a service in a different way (while maintaining the same interface and functionality) or by deploying the same implementation of a service as many standalone instances on virtual machines which differ in computational power. Such possible diversification in non-functional parameters of a service allows the service composition process to select services for participation in the workflow also on the basis of non-functional parameters. The process of service composition which considers end-to-end non-functional requirements for the composite service is referenced as Quality of Service (QoS) Aware Service Composition.

The QoS-Aware Service Composition enables the delivery of services that comply to user's non functional requirements. However, in many works the process of QoS-aware service composition is considered in scope of a single user request and uses constant values of service's non-functional parameters with the assumption that they are an average values of these parameters or that they are simply correct and up to date. Such assumption may be proper at the time of designing the service, yet if the service-oriented systems offers composite service that should comply with Service Level Agreement (SLA), such approach would not be sufficient. In order to be able to offer composite services which comply with SLA, the process of service selection for the composite service workflow should be performed at the arrival of composite service execution request and it should be preceded by thorough analysis of the system and requests processed by the system at the time of new request arrival in order to estimate the values of non-functional parameters of atomic services which may be selected to participate in the requested composite service. Proper service selection at this stage also enables the possibility of implementing a dynamic resource allocation mechanism, because services participating in the composite service may be selected in such way that they both, comply with non-functional requirements defined in the SLA for single service and optimize the resource allocation for the whole service-oriented system serving multiple requests at the same time.

While many research papers covered the topic of service composition, not much attention was put to the problem of both, fulfillment of user requirements and optimal resource allocation. This paper analyzes the mentioned problem and proposes an architecture of middleware for service-oriented systems, which would enable dynamic resource management with the consideration of users' non functional requirements, along with an optimization model for this problem. The article is organized as followed. Next section describes the motivation for the presented research. Section 3 briefly summarizes other research works in the field of QoS-Aware service composition and resource allocation. Section 4 describes the general architecture of complex service-oriented systems. Section 5 presents the proposed architecture of a the resource management middleware for service-oriented systems and request handling process implemented in this architecture. Finally, Sect. 6 formulates the optimization problem and proposes a model for the defined optimization problem.

2 Motivation

This research work is mainly motivated by the increasing importance of Quality of Service and resource management in service-oriented systems. It is also connected with development and constant increase in number of Internet of Things (IoT) systems, which often use web services for data acquisition, integration and processing. Due to a potentially large number of devices in such systems, the problem of proper and cost-effective resource allocation connected with quality of service assurance becomes a very important issue.

Analysis of many different research papers have also shown, that many of researches in this field focused on single elements of the problem without the consideration of the problem of quality of service management and resource allocation as a complete problem. Hence, in many works some of the elements of the problem were simplified, which lead to solution that where effective, but under some assumptions. This means that many of the works were only a part of the solution, which could be viable for implementation in production environments.

Many of the works also focused on the problem from the client perspective and tried to optimized the composite service in such a way, to optimize the QoS parameter values perceived by users and neglected costs generated for the service provider.

This work tries to address all the mentioned issues and proposes an architecture for service-oriented system along with the optimization problem formulation and model, which describes the complete problem of resource allocation and quality of service management in the context of proposed system architecture. This work is also considering the dynamic nature of service-oriented system and proposes a model for description of service's non-functional parameters values on the basis of their load, availability of communication resources and the characteristic of input data which is send to a web service.

3 Related Works

Problem of QoS-Aware service composition a topic of many research works. However, many of this works focused mainly on fulfillment of user requirements without the consideration of the resource allocation problem. Many of these works also select the web services for composite service workflow at design time when exact values of non-functional parameters are unknown [1–9]. These works often base their optimization algorithms only on expected or maximum values of non-functional parameters. Despite the value of this approach, a method for dynamic, performed at runtime, QoS-aware service composition is needed if we want to solve the combined problem of resource management and quality of service assurance.

One of the first papers discussing the problem of dynamic service composition described the eFlow composite service system and its features [10] like dynamic service discovery or dynamic conversation. However this work only considered the dynamic composition problem from the perspective of changes in user's functional requirements, without the consideration of service's QoS parameter values. Paper [11] analyzed the problem of e-health service personalization with the utilization of service composition. That paper introduced the concept of various service versions, which offered the same functionality, but differed in values of non-functional parameters. That paper also emphasized that the quality of service is connected with the resource allocation. The paper proposed different approaches to personalization of services delivered to the end-user, which were connected with the problem of resource allocation, however the proposed approach considered a limited set of parameters, hence it could not have be considered as full solution for the problem.

The importance of service quality was widely discussed in [12]. Authors indicated that composite service execution plans should consider possibility of changes in non-functional parameters values caused by delays, technical conditions, exceptions, failures, etc. in order to sustain the initially required service's QoS parameters. This work proposed a probabilistic model for service non-functional parameters description and computation. Such approach allows to model the dynamic characteristic of non-functional parameters. Other papers [13, 14] also notice the dynamic and uncertain nature of non-functional parameters' values. The benefit of such model is that, on average, selected services can perform better than services selected without knowledge about non-functional parameters' variability. However, probabilistic modeling still cannot guarantee better effectiveness than runtime non-functional parameter values estimation and does contribute to the solution of optimal resource allocation problem.

Finally, results from [12] were a base for works presented in [15, 16], which proposed a QoS-aware service composition approach, incorporating late binding, proxy services and execution monitoring, which enabled the possibility of dynamic service composition. The proposed approaches were based on constant monitoring of web-services and re-estimation of their non-functional parameters values. However, it is worth noting that this solution did not consider the problem of optimal research allocation itself, but it only focused on execution of composite service in such a way, which would guarantee the fulfillment of users non-functional requirements.

4 General Architecture of Complex Service-Oriented System

In order to propose a middleware for complex service-oriented systems, which will support the quality of service and resource management, it is crucial to analyze the general architecture of such systems. Despite the numerous solutions and patterns related to development of service-oriented systems, the main architectural designs are common for majority of them. Generally in service-oriented systems four main layers can be distinguished: service client layer, composite services layer, atomic services layer and infrastructure layer (Fig. 1).

The service client layer consists of all the applications invoking web-services offered by service oriented systems on behalf of their end-users. It is worth noting that applications, who are clients from this perspective, are not aware, if they are invoking a composite or an atomic service. From their perspective it is just a web service, which should be compliant with the contract (which includes both

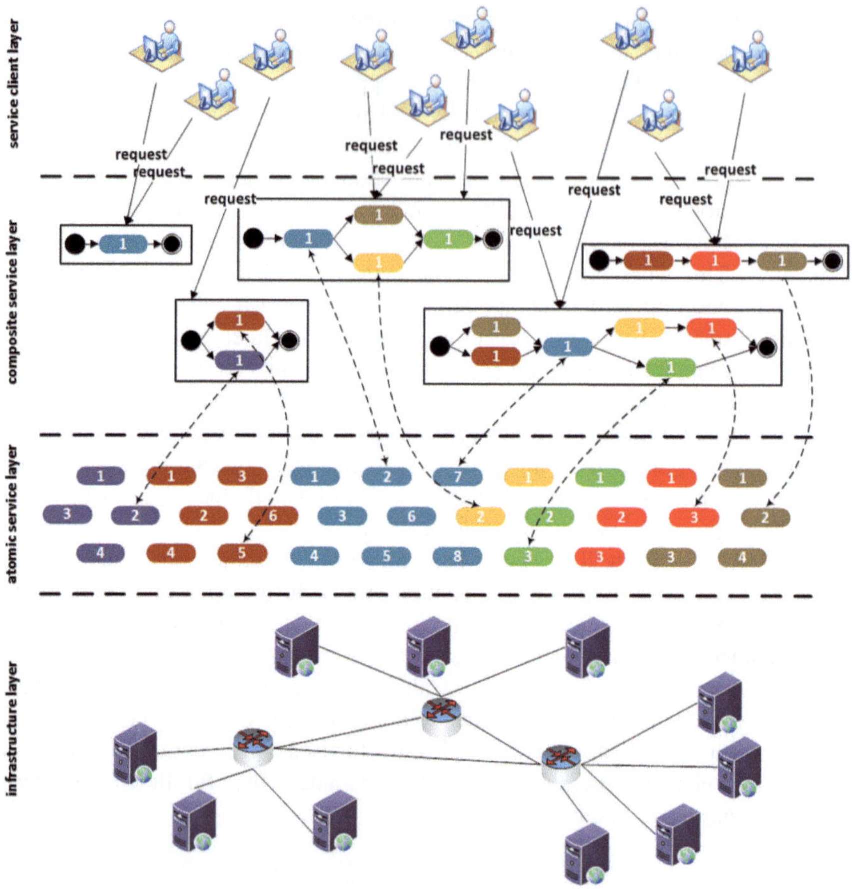

Fig. 1 General architecture of complex service-oriented system

functional and non-functional parameters). Also in many practical solutions each client has a dedicated interface for the composite web-service, which is bound to a specified composite service workflow and definition of non-functional requirements complaint with the SLA from the contract.

Second layer, the composite services layer, consists of all composite services offered by the system. This layer defines the complex functionalities offered by systems and it utilizes atomic services from the lower layer to deliver them. It is worth noting that such system can offer three types of composite services: a concrete composite service with all concrete atomic services participating in a workflow selected at design time, an abstract composite service for which concrete atomic services participating in a workflow are selected at runtime or a mixed services which contains both abstract and concrete atomic services. For the task of dynamic resource and quality management composite services in service-oriented system should be described as abstract services and should implement late binding mechanisms in order to enable service selection at runtime.

Third layer is the atomic services layer. In this layer all atomic services available in the system can be found. It is important to notice that some of the services can be available in different versions (which offer the same functionality, but differ in non-functional parameters). On the diagram different services are depicted by colors and different service versions are distinguished by numbers. It is worth noting that one atomic service can participate in numerous composite services at the same time, therefore the actual load of atomic service can influence the non-functional parameters of many composite services, in which it participates. From this perspective, proper allocation of atomic services to composite services is crucial for sustaining required level of quality. Such allocation of services to composite services also directly corresponds to computational resource allocation, because in majority of scenarios atomic services are deployed in virtualized environments where services are deployed on dedicated virtual machines, which computational power can be fully utilized by the service. Even in scenarios where multiple services are deployed on single virtual machine or servers the same isolation of resources can be achieved, because web application servers (for instance, Internet Information Services in Microsoft Windows Server) often offer mechanisms for limiting resources like processor time or memory allocation for each single web service deployed on them.

The last layer is the infrastructure layer. This layer represents all the hardware on which atomic services are deployed. In modern service-oriented systems hardware infrastructure is often widely distributed or it is even based on cloud-based platforms offered by third-party companies. From the perspective of resource allocation, this layer can be used to manage computational resources of each atomic service and communication resources available for the service-based system. The approach presented in this paper assumes that middleware system cannot directly control this layer (because infrastructure may be managed by other authority, e.g. cloud platform provider, or because middleware system may use services that are offered by external organizations), hence the resource allocation is conducted by dynamic binding of available atomic services (with their own dedicated resources) to composite services which are being executed in the service-oriented system. Also, in this approach, each utilization of atomic service is considered as a cost incurred by the service provider.

5 Dynamic Resource Management Middleware Architecture

The proposed quality of service and resource management middleware is a part of the composite service layer described in the general architecture of complex service-oriented system. The diagram (Fig. 2) presents platform independent model

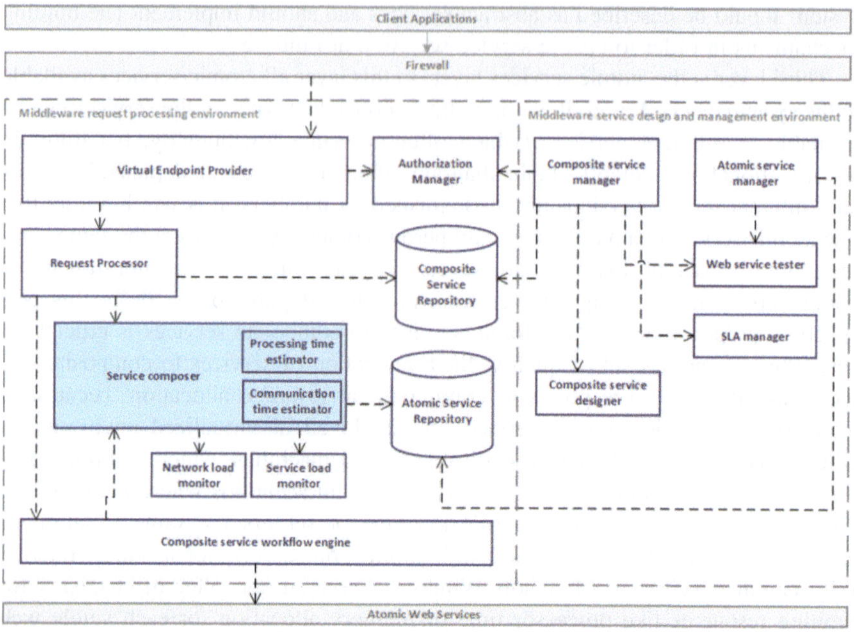

Fig. 2 Dynamic resource management middleware architecture diagram

of the middleware system, where all system components and their dependencies were depicted.

It is worth noting that the final implementation of the system may differ (e.g. some of the components may be merged due to performance requirements), however all functions offered by described components must be preserved in order to maintain the full functionality of the proposed middleware system. The architecture of the middleware can be divided into two main parts: the middleware request processing environment and the middleware service design and management environment.

The *service design and management environment* consists of all the components which help in service registration, management and testing. *Atomic Service Manager* allows service provider to register atomic services in the atomic service repository. It also allows to define their non-functional characteristics on the basis of testing and request processing patterns (one request at a time with a queue, multiple requests at one time without a queue, multiple requests at one time with an additional queue) which are stored in atomic service repository and use for the estimation of atomic service processing time along with other non-functional parameters. Composite service manager allows the service provider to design composite service workflows (abstract, concrete or mixed) with the use of a dedicated composite service design interface. Service provider can also manage the security of such composite service with the use of *Authorization Manager*, he can manage its SLA with *SLA Manager* and he can publish dedicated virtual endpoints for service clients with the use of *Virtual Endpoint Provider*. On the whole, the service design and management environment delivers tools that help to manage and properly describe the non-functional characteristics of web services on the basis of testing, which are crucial for proper resource and quality of service management during composite service workflow execution.

The *service request processing environment* is responsible for dynamic quality of service and resource management during the execution of the user requests. The request handling process begins in the *Virtual Endpoint Provider*. After receiving the service execution request *Virtual Endpoint Provider* verifies client's privileges with the help of *Authorization Manager*.

On the basis of this verification request is rejected or it is forwarded to the *Request Processor*. The *Request Processor* is the main component responsible for the management of requests life-cycle. First it downloads the composite service workflow description along with the SLA contract, which are connected with the processed request and the service client. After downloading the service description and SLA, *Request Processor* sends them to *Service Composer*.

Service Composer is the main component responsible for quality of service and optimal resource allocation in service-oriented system and it implements optimization algorithms build on the basis of the model proposed in the Sect. 6. The service composer first finds the matching services in the *Atomic Service Repository* and downloads their non-functional characteristics. Next, the *Processing Time Estimator* and *Communication Time Estimator e*stimate processing time of each atomic services which can participate in the composite service workflow and the data transfer time between services on the basis of service characteristics and current load data from *Service Load Monitor* and *Network load monitor*. On the basis of these estimations, *Service Composer* searches for most cost-effective set of services which can participate in the composite service workflow. After finding the services, the composite service workflow with concrete atomic services is sent back to the *Request Processor*. If the service composition was successful *Request Processor* forwards the composite service execution plan to the *Composite Service Workflow Engine* and if *Service Composer* was unable to find a service execution plan which complies with the required SLA the service execution request is reject by the *Request Processor*. After receiving the composite service execution plan *Composite Service Workflow Engine* begins its execution and monitors the workflow. If the workflow execution is not complaint with the execution plan, the *Composite Service Workflow Engine* can request the *Service Composer* for dynamic re-composition of the composite service workflow in order to sustain required non-functional parameters of the composite service. After the successful execution of the composite service, *Composite Service Workflow Engine* returns the result to the *Request Processor*, which returns the result to the service client through a dedicated virtual service endpoint. It is worth mentioning that such modular architecture enables scalability of the systems, because each of the components may be deployed in many independent instances, hence this middleware would be also able to work in environments with very high request rates.

6 Resource Allocation and Quality of Service Management Problem

The problem of resource allocation and quality of service management is a combined problem of constraint satisfaction and optimization. In this interpretation constraint satisfaction is related to the problem of SLA compliance. The service offered by the service provider must meet the defined values of non-functional parameters in order to properly handle client requests and avoid penalties. On the other hand, is also an optimization problem because the most important metric for the service provider is the profit, hence he must minimize his service delivery costs while sustaining the required service quality in order to maximize the profit.

In order to solve the stated problem, a model which would consider the dynamic characteristic of non-functional in the process of resource allocation optimization and quality of service management is proposed in this section. The model is based on the architecture model presented in previous sections and may be practically utilized in the development of optimization method for Service Composer component. Proper modeling of the dynamic changes in services is crucial for the accuracy of the optimization model. The following paragraphs define a proposition of such representations.

Let's define the required non-functional parameters for the request k arriving at moment t as a vector:

$$R_k = [r_{k1}, r_{k2}, \ldots, r_{ki}, \ldots, r_{kI}] \tag{1}$$

where i is the index of non-functional parameter and I is the total number of non-functional parameters.

Also, let's define the set of all available atomic services as:

$$S = \{s_1, s_2, \ldots, s_j, \ldots, s_J\} \tag{2}$$

where j is the index of the atomic service s and J is the total number of services.

Following, lets define the load of the services as a vector:

$$L_t = [l_{t1}, l_{t2}, \ldots, l_{tj}, \ldots, l_{tJ}] \tag{3}$$

where l_{tj} is the number of requests processed by the service j at the moment t.

Let's define communication capacity available in the system as matrix:

$$C_t = [c_{tij}] \tag{4}$$

where c_{tij} is the available network capacity of the network link between services s_i and s_j at the moment t.

Also, let's define the data size which is sent with the request as d_k and d_{kij}, where d_k describes the size of data sent with the request k, and d_{kij} describes the size of the data sent between services s_j and s_i participating in the workflow of composite service realizing request k.

The graph representing all possible realizations of composite service workflow for request k can be defined as graph:

$$GS_k(VS_k = \{V_{START}, VC_1, VC_2, \ldots, VC_n, V_{END}\}, ES_k) \tag{5}$$

where $VC_n = \{s_{n1}, s_{n2}, \ldots, s_{nl}\}$ is a set of all web services complaint with functional requirements for n-th service participating in composite service workflow, V_{START}, V_{END}, represent start and end node of the composite service workflow. VS_k is a set of all vertexes of the graph, which represent all abstract services participating in the workflow along with the set of all candidate services complaint with functional requirements and ES_k, which is a set of all edges of the graph, represents all data-flow connections between abstract web services.

Following the composite service execution workflow for request k can be defined as graph:

$$G_k(V_k, E_k) \tag{6}$$

where V_k represents all concrete web services participating in the workflow and E_k describes all data-flow connections between concrete web services. On this basis we can define:

$$G_{kf}\left(V_{kf}, E_{kf}\right) \in FG_k(V_k, E_k) \tag{7}$$

where $FG_k(V_k, E_k)$ is a set of all possible composite service execution workflows for request k build on the basis of graph GS_k and $G_{kf}\left(V_{kf}, E_{kf}\right)$ is f-th composite service execution workflow for request k from the set FG_k.

For each non-functional requirement r_i and composite service workflow graph G_{kf} we can defined non-functional parameter aggregation function as:

$$f_{r_i}\left(G_{kf}\left(V_{kf}, E_{kf}\right), L_t, C_t, d_k\right) \tag{8}$$

From the perspective of service provider, we can define composite service k delivery cost as:

$$DC_{tk}\left(G_{kf}\left(V_{kf}, E_{kf}\right), L_t, C_t, d_k, R_k, p_k\right) \tag{9}$$

where p_k represents penalty for not fulfilling non-functional requirements defined in SLA for request k.

Finally we can define optimization problem as:

$$G_k(V_k, E_k) \leftarrow \min_{G_k(V_{kf}, E_{kf})} DC_{tk}\left(G_{kf}\left(V_{kf}, E_{kf}\right), L_t, C_t, d_k, R_k, p_k\right) \quad (10)$$

subjected to:

$$f_{r_i}\left(G_{kf}\left(V_{kf}, E_{kf}\right), L_t, C_t, d_k\right) < r_i \quad (11)$$

It is worth noting that DC function definition should be declared individually for each service-oriented system, according to its business model, because they can use different atomic service hosting methods, e.g. some can have their private hardware infrastructure and they have to calculate service execution cost on the basis of the hardware infrastructure maintenance cost, while other can use infrastructure provided by cloud infrastructure providers, which often offer pay-per-use model and can scale on demand in order to provide better non-functional parameters for additional fee.

7 Conclusions

This article describes the combined problem of quality of service management and optimal resource allocation in service-oriented systems and describes the general architecture of complex service oriented systems. On the basis of this analysis and description, this paper proposes an architecture for dynamic resource management middleware with the description of the service handling process performed by the middleware build according to this architecture. The proposition of the architecture is followed by the definition of resource allocation and quality of service management problem and a proposition of a optimization model for this problem, which matches the described middleware system architecture. Further research works will focus on the implementation of a test complex service-oriented system and middleware build according to the proposed architecture, which will be used for verification of the optimization model and optimization methods build according to the proposed model. Later, the research will be fully dedicated to model the non-functional aggregation functions and service provider cost functions for different types of service-oriented systems, followed by the development of effective and efficient optimization methods, which would be viable for dynamic quality of service management and resource allocation optimization.

Acknowledgements The research presented in this paper was partially supported by the Polish Ministry of Science and Higher Education and the European Union within the European Regional Development Fund, Grant No. POIG.01.03.01-02-079/12 and within European Social Fund.

References

1. Cao, H., Feng, X., Sun, Y., Zhang, Z., Wu, Q.: A service selection model with multiple QoS constraints on the MMKP. In: Network and Parallel Computing Workshops, pp. 584–589 (2007)
2. Bao, H., Dou, W.: A QoS-aware service selection method for cloud service composition. In: Parallel and Distributed Processing Symposium Workshops & PhD Forum (IPDPSW), pp. 2254–2261 (2012)
3. Stelmach, P., Grzech, A., Juszczyszyn, K.: A model for automated service composition system in SOA environment. In: Technological Innovation for Sustainability. Springer, Heidelberg, cop. 2011, pp. 75–82 (2011)
4. Blanco, E., Cardinale, Y., Vidal, M., Graterol, J.: Techniques to produce optimal web service compositions. In: IEEE Congress on Services, pp. 553–558 (2008)
5. Strunk, A.: QoS-Aware service composition: A survey. In: Eighth IEEE European Conference on Web Services, pp. 67–74 (2010)
6. Huang, A.F.M., Lan, C., Yang, S.J.H.: An optimal QoS-based web service selection scheme. Inf. Sci. **179**(19), 3309–3322 (2009)
7. Świątek, P., Stelmach, P., Prusiewicz, A., Juszczyszyn, K.: Service composition in knowledge-based SOA systems. In: New Generation Computing, pp. 165–188 (2012)
8. Juszczyszyn, K., Stelmach, P., Grzelak, T.: A method for the composition of semantically described Web services. In: Information Systems Architecture and Technology: Networks and Networks' Services, pp. 27–37 (2010)
9. Stelmach, P., Falas, L.: Service composer: A framework for elastic service composition. Information systems architecture and technology. In: Service Oriented Networked Systems, pp. 47–60 (2011)
10. Casati, F., Shan, M.C.: Dynamic and adaptive composition of eservices. Inf. Syst. **26**(3), 143–163 (2001)
11. Grzech, A., S´wia̡tek, P., Rygielski, P.: Dynamic resources allocation for delivery of personalized services. In: Software Services for e-World, Volume 341 of IFIP Advances in Information and Communication Technology. Springer, Berlin, pp. 17–28 (2010)
12. Cardoso, A.J.S.: Quality of service and semantic composition of workflows. Dissertation, University of Georgia (2002)
13. Hwang, S.Y., Wang, H., Tang, J., Srivastava, J.: A probabilistic approach to modeling and estimating the QoS of web-services-based workflows. Inf. Sci. **177**, 5484–5503 (2007)
14. Falas, L., Stelmach, P.: Web service composition with uncertain nonfunctional parameters. In: Technological Innovation for the Internet of Things. Springer, Berlin, pp. 45–52 (2013)
15. Canfora, G., Di Penta, M., Esposito, R., Villani, M.L.: Qos-aware replanning of composite web services. In: Proceedings of Web Services, ICWS 2005 (2005)
16. Canfora, G., Di Penta, M., Esposito, R., Villani, M.L.: A lightweight approach for QoS-aware service composition. In: Proceedings of 2nd International Conference on Service Oriented Computing ICSOC'04 (2004)

Part II
High Performance and Cloud Computing

Method of Solving the Volume Between Triangulated Mesh Surfaces

Szymon Tengler and David Sehnal

Abstract Surfaces built out of triangles are often used for computer modeling of terrain unevenness. A well known and studied type of such surface is the Triangulated Irregular Network (TIN). TIN surfaces are often used in geographical information systems (GIS) for modeling coal deposits or in oceanography. In such cases the terrain is usually split into layers modeled as the TIN surfaces. In order to find the mass of a particular terrain substance (e.g. coal) it is required to compute the signed volume bounded by two TIN surfaces. The article presents a comprehensive algorithm for solving this problem, including a reference implementation in C# language.

Keywords Triangulated surfaces · Surface volume calculation · Algorithm

1 Introduction

This article is a response to a real demand for solving the problem of calculating the volume between two triangulated surfaces—TIN [1]. Several tools for solving this kind of problem are available (e.g. Microstation [2], ArcGIS [3]). However, most of them are commercial, and their purchase is relatively expensive. Moreover, their

S. Tengler (✉)
Faculty of Mechanics and Informatics, University of Bielsko-Biala,
Bielsko-Biala, Poland
e-mail: stengler@ath.bielsko.pl

D. Sehnal
Faculty of Informatics, Masaryk University Brno, Brno, Czech Republic
e-mail: david.sehnal@mail.muni.cz

© Springer International Publishing Switzerland 2016

J. Świątek et al. (eds.), *Information Systems Architecture and Technology: Proceedings of 36th International Conference on Information Systems Architecture and Technology – ISAT 2015 – Part III*,
Advances in Intelligent Systems and Computing 431, DOI 10.1007/978-3-319-28564-1_5

main drawback is that their extension and cooperation with other software required in TIN analysis is strictly determined by the producer of the solution.

According to the authors of the available software, performing the calculation of the volume is relatively difficult. Manufacturers do not disclose the details on how to perform this type of calculation. Available publications include only some of the most common issues that are included in the developed algorithm—e.g. triangulation, locating a point in a triangulation, or polygon intersection. The algorithm proposed in this article is a complete solution to calculating the volume between two triangulated surfaces, which allows implementation in any programming language—e.g. C/C++, C#, or Python. The article includes a reference implementation written in C#.

2 Definition of the Problem

Given two TIN surfaces, A and B, we are interesting in computing signed volume defined by the differences in surface elevation on the Z axis (i.e. over the XY plane). The surfaces between which the volume is sought can be positioned differently in relation to each other. Two general cases—non-intersecting surfaces and intersecting surfaces—are presented in Fig. 1.

These surfaces can be of any size and oriented in any way in relation to each other (excluding the situation when they are mutually perpendicular). In the case of the intersecting surfaces (Fig. 1b) the volume with the plus or minus sign depending whether surface *A* is "under" or "over" surface *B*, is sought.

The volume in the area being an overlapping part of the edges of two surfaces in question is sought in the presented cases—this area is marked in Fig. 2.

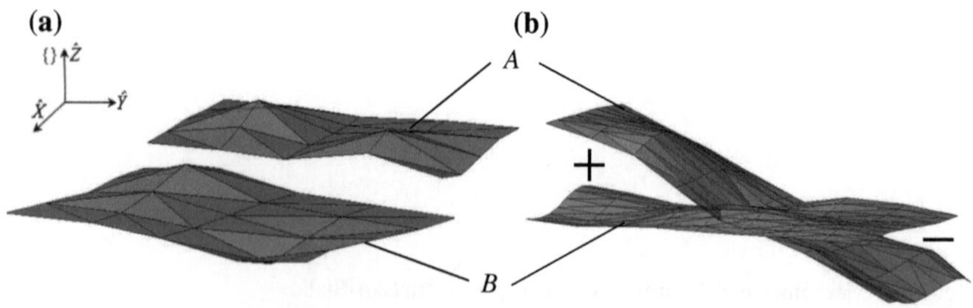

Fig. 1 Two surfaces: **a** non-intersecting, **b** intersecting

The area is defined by the overlapping part of polygons constituting edge outlines of surfaces *A* and *B* projected onto plane *XY*. The outline of the obtained area can be a concave or convex polygon depending on the shape of the input surfaces.

It is assumed that the area in which the volume is calculated can additionally be limited by another arbitrary boundary (defined as a convex or concave polygon)—Fig. 3.

In such a case the volume is calculated over the area defined by the overlapping parts of three elements: outlines of surfaces *A* and *B* projected onto plane *XY*, and the additional limiting boundary—Fig. 4. We refer to the yellow region as a "connected component".

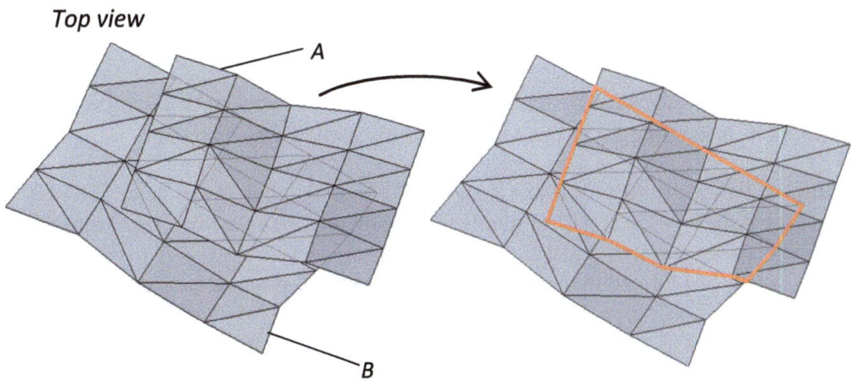

Fig. 2 Two non-intersecting surfaces with marking their overlapping part from a top view

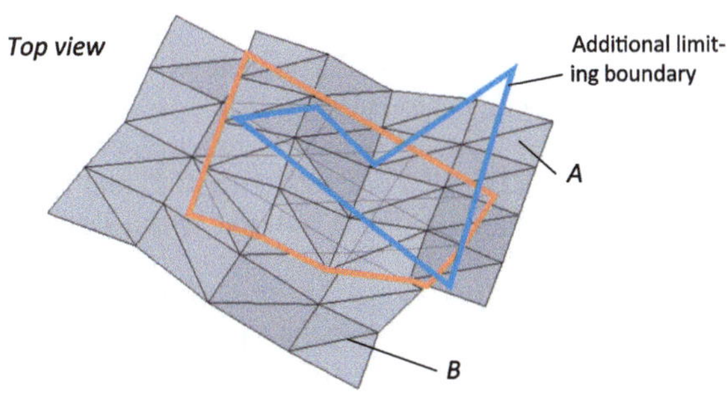

Fig. 3 Limitation of the boundary in which the volume is calculated

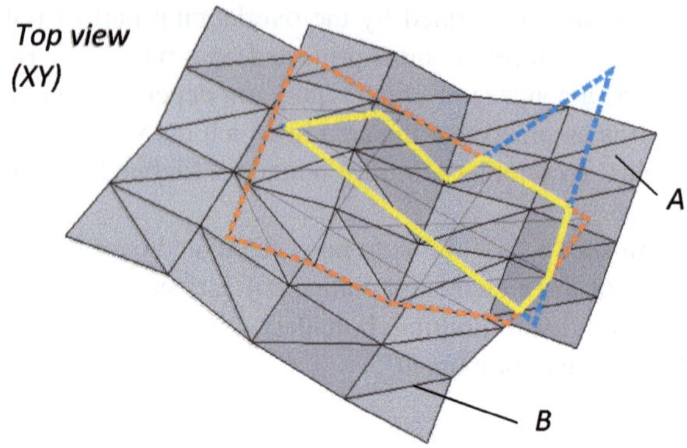

Fig. 4 The area (marked in *yellow*), in which the volume is calculated after implementing the additional limiting area

3 Methods

This section describes the basic steps of the algorithm. A more detailed description is available in comments in the reference implementation.

Input: Triangulated 3D surfaces A and B, 2D bounding polygon P on the XY plane
Output: (V^+, V^-) - a pair of numbers representing the "positive" and "negative" volume between surfaces A and B over the polygon P.

1. Q - polygon formed by intersecting 2D boundary of A and B projected to the XY plane and P
2. (A', B') - clip A and B on the XY plane by Q
3. (V^+, V^-) - initialize negative and positive volumes to 0
4. For each pair of connected components of A' and B':
   a) $S \leftarrow$empty stack, $H \leftarrow$ empty set of visited pairs
   b) Find a pair of triangles $(T_A \in A', T_B \in B')$ that overlap on the XY plane and push it to S. If no such pair exists, skip this pair.
   c) While S is not empty:
      - $(T_A, T_B) \leftarrow$ pop S
      - If not H contains (T_A, T_B) then:
      - Add (T_A, T_B) to H
      - $K \leftarrow$ 2D region of intersection of T_A and T_B where plane of T_A is above T_B
      - $L \leftarrow$ 2D region of intersection of T_A and T_B where plane of T_A is bellow T_B
      - $(V_K^+, V_L^-) \leftarrow$ volume between T_A and T_B over K, volume between T_A and T_B over L
      - $(V^+, V^-) \leftarrow (V^+ + V_K^+, V^- + V_L^-)$
      - If $V_K^+ > 0$ or $V_L^- > 0$ then push all pairs of neighbors of T_A and T_B to S
5. Return (V^+, V^-)

Steps 1 and 2 are a straightforward and a well established [4, 5].

In step 4 it is necessary to identify connected components of A′ and B′ because in the general case, after the clipping by P, the surfaces might not remain connected as illustrated in Fig. 5.

Computing the 2D region of intersection (Fig. 6) of two triangles is also a straightforward, albeit tedious, process.

Complexity

The complexity of the algorithm is $O(E_m log E_m + MN)$ where E_m is the maximum number of edges of 2D boundary of A, B and the polygon P; M, N is the number of triangles in A and B respectively. The $O(E_m log E_m)$ part corresponds to the complexity of step 1. Each iteration of step 2 takes $O(M + N)$ steps to finish. In step 4 the worst case complexity is $O(MN)$ in the case the identification of two intersecting triangles needs to check all the possible pairs. However, this worst case scenario happens only in very degenerate cases of choosing the polygon P. In most practical cases, the complexity of the 4th step can be expected to be $O(M + N)$ due

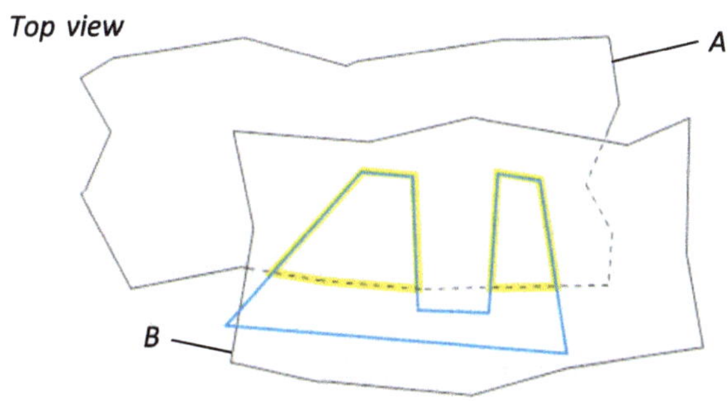

Fig. 5 Sketch presenting connected components marked in *yellow*

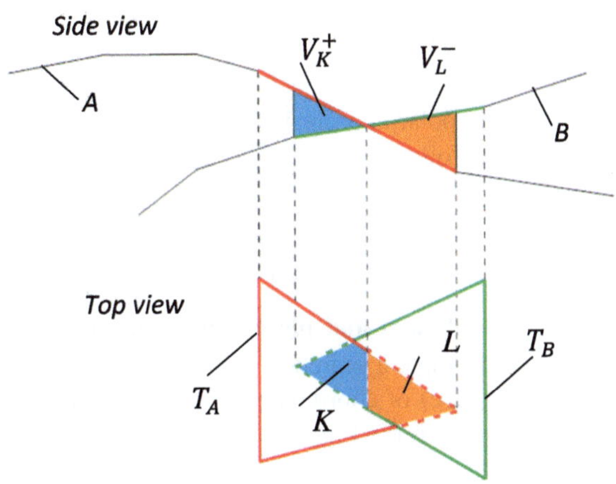

Fig. 6 Intersection of two triangles T_A and T_B

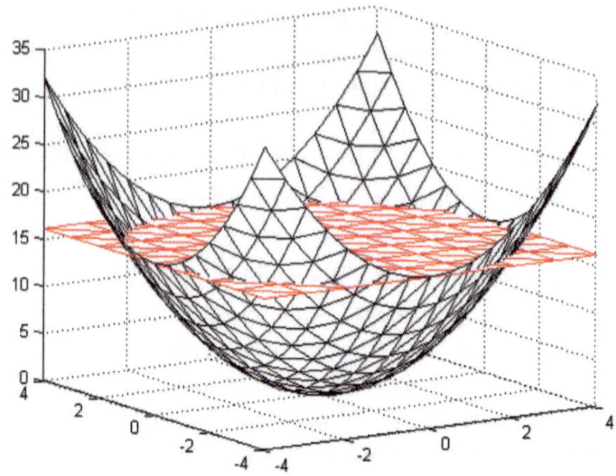

Fig. 7 Searched volume limited by paraboloid and plane

to the fact that each triangle from A' will be over some triangle from B'. As such, the expected running time of the algorithm is $O(E_m log E_m + M + N)$.

4 Testing

Five examples of volume calculation and an analysis of the algorithm's accuracy and speed of the reference implementation are presented. After that the results comparison are presented.

Test 1—paraboloid and plane
In the first example, the sought volume is limited by surfaces: $z_1 = x^2 + y^2$ and $z_2 = 16$—the surfaces are illustrated on the Fig. 7.

Analytical solutions can by calculated as:
$V^+ = \iint_D 16dA - \iint_D (x^2 + y^2)dA = \iint_D 16 - (x^2 + y^2)dA$, where $D = \{(x, y) : x^2 + y^2 = z^2\}$.

Using the polar coordinates: $x = r cos\theta, y = r sin\theta$, the value of integral can be calculates as follows:

$$\int_0^{2\pi} \int_0^4 r(16 - r^2) dr d\theta = 128\pi.$$

Test 2—paraboloid and plane limited by polygon
In second example a searched volume is limited by $z_1 = x^2 + y^2$, $z_2 = 16$ and the polygon (square) of coordinates $(-1, -1, 25), (1, -1, 25), (1, 1, 25), (-1, 1, 25)$—it is illustrated on the Fig. 8.

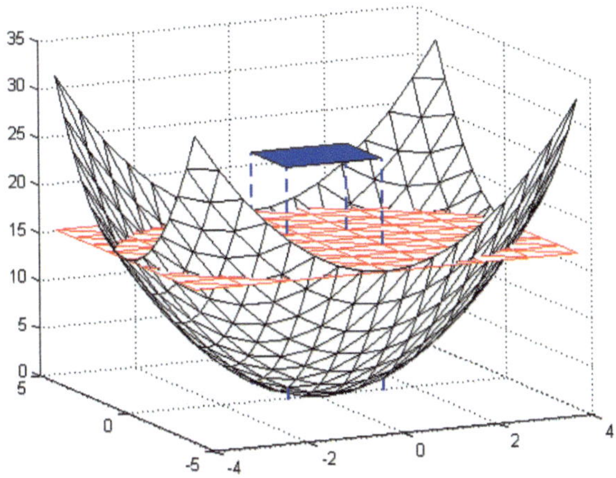

Fig. 8 Searched volume limited by: paraboloid, plane and polygon

It this case the limited volume can be calculated as:

$V^+ = 64 - \iint_D (x^2 + y^2)dA, \quad \text{where} \quad D = \{(x, y) : -1 \leq x \leq 1; -1 \leq y \leq 1\}.$

Taking the integrated region D into account:

$$V^+ = 64 - \int_{-1}^{1}\int_{-1}^{1} \left(x^2 + y^2\right) dxdy = 61\frac{1}{3}.$$

Test 3—sinusoid surface and plane

In third example, searched volume is limited by $z_1 = \sin x \sin y$ where $0 < x < 2\pi$, $0 < y < 2\pi$ and $z_2 = 0$—it is illustrated on the Fig. 9.

In this case volume with signs V^+ and V^- is expected. Surface z_1 is symmetrically to $z_2 = 0$, so each of them can by calculated by:

$V^{\pm} = \pm\frac{1}{2}\iint_D |\sin x \sin y| dA$, where $D = \{(x, y) : 0 \leq x \leq 2\pi; 0 \leq y \leq 2\pi\}$.

The result of integral is: $V^{\pm} = \pm\frac{1}{2}\int_0^{2\pi}\int_0^{2\pi} \sin x \sin y \, dxdy = \pm 8$.

Test 4—sinusoid surface and plane limited by polygon

In the next example a searched volume is limited by $z_1 = x^2 + y^2$, $z_2 = 16$ and polygon (square) of coordinates $\left(\frac{\pi}{2}, \frac{\pi}{2}\right)$, $\left(\pi, \frac{\pi}{2}\right)$, (π, π), $\left(\frac{\pi}{2}, \pi\right)$ – it is illustrated on the Fig. 10.

The limited volume can be calculated as:

$V^+ = \iint\limits_D \sin x \sin y \, dA$, where $D = \left\{(x, y) \colon \frac{\pi}{2} \leq x \leq \pi; \frac{\pi}{2} \leq y \leq \pi\right\}$.

Taking the integrated region D into account:

$$V^+ = \int_{\frac{\pi}{2}}^{\pi}\int_{\frac{\pi}{2}}^{\pi} \sin x \sin y \, dxdy = 1.$$

Fig. 9 Searched volume limited by sinusoid and plane

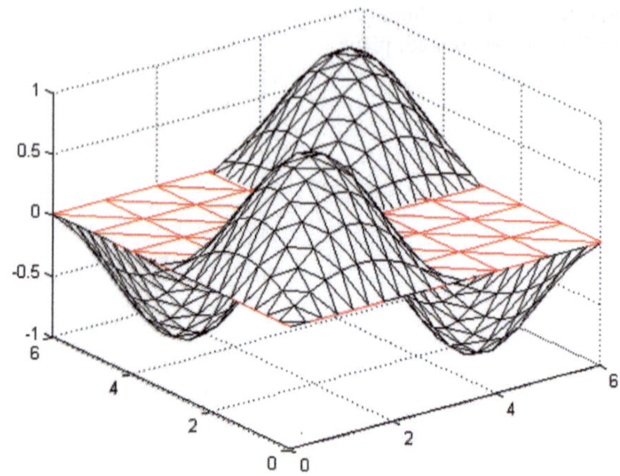

Fig. 10 Searched volume limited by: sinusoid, plane and polygon

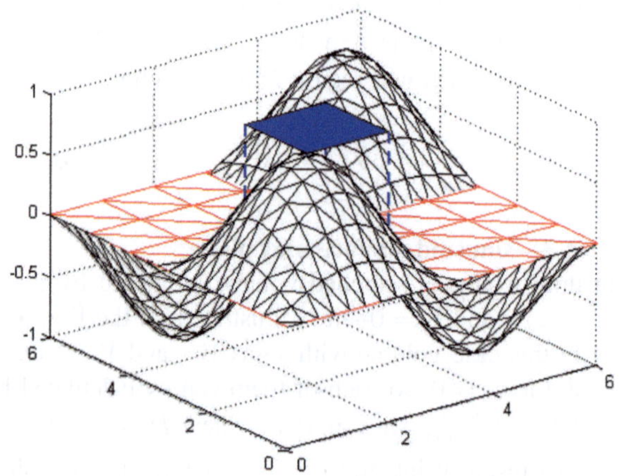

Test 5—two surfaces limited by polygon

The last one example presents two surfaces $z_1 = y^2 - x^2 + 10$, $z_2 = xy + 40$ and limiting triangle of coordinates $(-3, -3), (3, -3), (-3, 3)$—Fig. 11.

The limited volume can be calculated as:

$V^+ = \iint_D((xy+40) - (y^2 - x^2))dA$, where $D = \{(x, y){:} -3 \leq x \leq 3; -3 \leq y \leq -x\}$. Taking the integrated region D into account:

$$V^+ = \int_{-3}^{3} \int_{-3}^{-x} \left((xy+40) - \left(y^2 - x^2\right)\right) dxdy = 540.$$

The comparison of results calculated both analytically and by our implementation of the algorithm is given in a Table 1. Results comparision.

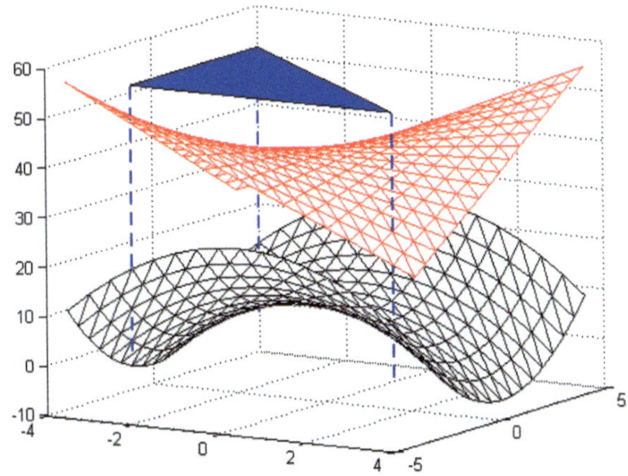

Fig. 11 Searched volume limited by: sinusoid, plane and triangle

Table 1 Results comparision

	Calculated analytically		Calculated by program		Relative error (%)
	V^+	V^-	V^+	V^-	
Test 1	128π	–	401.84	–	0.07
Test 2	$61\frac{1}{3}$	–	61.32	–	0.02
Test 3	8	−8	7.99	−7.99	0.13
Test 4	1	–	1	–	0
Test 5	540	–	+540.01	–	0.002

The relative error is calculated as:

$$E = \frac{|V^A - V^P|}{|V^A|} 100\ \%,$$

where

V^A volume calculated analytically,

V^P volume calculated by program.

The presented accuracy of calculation depends on number of triangles in particular surface. More triangles improves the accuracy but the computation time gets longer. The calculation time and calculation error relative to number of triangles are shown in Fig. 12.

The tests were performed on a computer with Intel Core i7 2.2 GHz processor. The charts present that the error was stabilized at 800 triangles and at the calculation time 0.03 s.

In order to not complicate the analytical calculations in the presented examples, simple shapes of the limiting polygon were used—a rectangle and a triangle. As mentioned earlier in the article, the developed algorithm can accept any convex or concave polygon. However, because any 2D polygon can be triangulated, the presented case studies are representative.

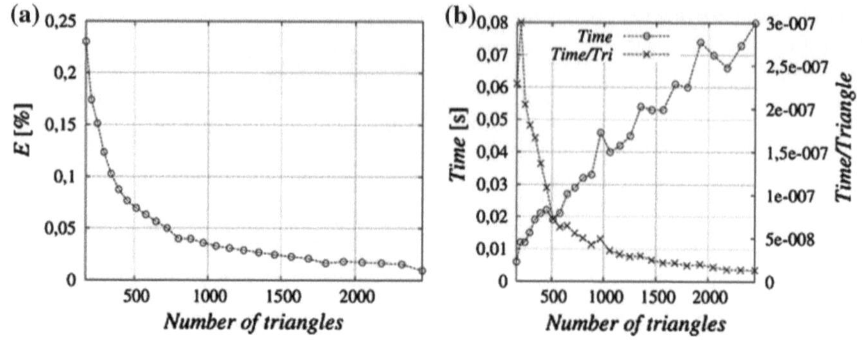

Fig. 12 Calculation: **a** calculation error, **b** time and time per triangle

5 Conclusions

This article presents an algorithm for calculating the volume between two triangulated surfaces. The surfaces can intersect each other and be limited by additional polygon of any shape. A reference C# implementation is also provided under the LGPL license. This implementation was verified on examples with known results. We have shown that the algorithm is fast and accurate for sufficiently dense triangulations. The developed solution is suitable for example in development of software used by geodesists.

References

1. Gold, C.: Digital Terrain modeling: Principles and methodology. Taylor and Francis, London (2007)
2. http://www.bentley.com/en-US/Products/microstation+product+line/
3. http://www.esri.com/software/arcgis
4. O'Rourke, J.: Computational geometry in C, (2nd ed.). Cambridge University Press, Cambridge. ISBN 0-521-64976-5 (1998)
5. Berg, M., Otfried, Ch., Kreveld, M., Overmars, M.: Computational Geometry: Algorithms and Applications, (3rd edn). ISBN: 978-3-540-77973-5 (2008)

Accelerated Additive Algorithm with Lossy Data Compression

Aleksandr Katkow

Abstract In this paper the accelerated algorithm of a calculation of Discrete Cosine Transform (DCT), widely used in the practice of the conversion of graphic images, is considered. We investigate the process of the calculating DCT by means of additive formulas. The application of this formulas allows to obtain the very fast computing schemes, which are so necessary for real time systems. In this article is discussed the algorithm, in which the computational process it is process of the algebraic summing basic function samples taken at certain points of the interval independent variable. The conception of the accelerated calculation of DCT is considered on example of real graphic images. The presented algorithms are oriented first of all on use for the analysis of discrete and analog signals and image processing in real time. Additive algorithm for image processing are presented in meta language.

Keywords High performance computing · Lossy data compression · Spectral analysis

1 Introduction

The task of constructing fast algorithms for spectral analysis is relevant and now it is characterized by a large number of publications on this topic. Considerable attention is paid to the construction of formulas that allow to obtain the value of the spectrum is not always accurately. In these publications it comes to some loss of accuracy to obtain a high and in some cases, very high speed processing of information. Significant part of the publications is devoted building new algorithms

A. Katkow (✉)
University of Computer Sciences and Skills, Lodz, Poland
e-mail: aleksandr.katkow@yahoo.com

© Springer International Publishing Switzerland 2016

J. Świątek et al. (eds.), *Information Systems Architecture and Technology: Proceedings of 36th International Conference on Information Systems Architecture and Technology – ISAT 2015 – Part III*, Advances in Intelligent Systems and Computing 431, DOI 10.1007/978-3-319-28564-1_6

for computing the spectrum [3, 4] or using a smaller number of arithmetic operations [1, 2], in another part of the works the number of operations is reduced by the fact that multiplication is eliminated. Instead, uses only addition. Such algorithms are called additive [5–8].

There are two embodiments of the additive algorithms for fast calculation of the Fourier integral using additive formulas, which are respectively called direct additive formulas and dual additive formulas [5]. When implementing the second embodiment of the additive algorithm for computing the Fourier integral it is done by means of the algebraic summation of basic functions, taken at the points, which are determined by calculating the definite integral function and comparing its value with a certain value called threshold. When the value of the definite integral of the analyzed function becomes equal to the threshold value at these points take place fixing the values of basic functions. These values of the basic functions algebraically summed. Thus in the implementation of dual additive formulas take place irregular grid of values of the independent variable and they is not known in advance, because they depend on the values of the analyzed function. This embodiment of the additive of formula is designed to implement directly into an analog-digital converter, and in that its part which works with the analog representation the analyzed function, and may for example consist of an analog integrator and comparator. In this case, the transformation of the analyzed function in digital form is generally not required, it is necessary only to have numerical value of the independent variable at those points where the value of the definite integral of analyzed function is equal to the predetermined threshold value. The basic algorithm of dual additive formulas of the Fourier transform has been proposed a long time [6]. Later, in [7], it was proposed to use this algorithm to compute the Fourier coefficients of the solution of differential equations in partial derivatives. In the same article, study was conducted filtering properties of additive algorithms and their implementation by means of direct and dual formulae. In the next publication [8] we consider using additive formulas for the calculation of the spectral density of the signal. In this paper we consider attempt to use additive dual formulas for calculation of DCT in image compression.

2 Preliminaries

Consider in more detail a method for producing dual formulas additive spectral analysis for the approximate calculation of Fourier integrals. If the basic part of energy of continual signals is concentrated in the limited intervals of time [−T/2, T/2] and frequencies [−Ω/2, Ω/2], then we have the following approximate relations between a signal F(t) and a spectrum S(ω) of this signal

$$S(\omega) = \int_{-T/2}^{T/2} F(t)\exp(-j\omega t)dt, \quad (1)$$

$$F(t) = 1/2\pi \int_{-\Omega/2}^{\Omega/2} S(\omega)\exp(j\omega t)d\omega. \quad (2)$$

where

$$S(\omega) = A(\omega) + jB(\omega)$$

$$F(t) = f(t) + j\varphi(t)$$

The essence a method of an approximate calculation of Fourier integrals using the dual formula is as follows. The technology of processing of information used in the construction of dual formulas is similar to technology of processing of information in neural networks. It is known that in neural networks individual neurons generate an output signal when the sum of input signals exceeds a certain threshold. Also, in the implementation of dual AFT formulas is proposed to integrate the investigated function and compare the value obtained by the integral of a function with a threshold. When the value of the integral of the input function equals the value of the threshold, it is proposed to fix value of the basic functions at these points. The obtained values of the basic functions are used in the dual formulas of AFT to calculate values of the spectral density of the input function. One can get the integrals (1, 2) using the algebraic addition of the values of $\sin \omega t$ and $\cos \omega t$, taken in a specific discrete values of time and frequency. We consider this method in detail. The whole interval of integration in (1) is divided into small intervals so that we could present the function $e^{-j\omega t}$ with a sufficient accuracy by linear terms of the Taylor series on intervals. At each interval of integration we have

$$\int_{t_u}^{t_{u+1}} F(t)e^{-j\omega t}dt = e^{-j\omega t_u} \int_{t_u}^{t_{u+1}} F(t)dt + \delta_u^S(\omega),$$

where

$$\delta_u^S(\omega) = \omega e^{-j\left(\frac{\pi}{2} + \omega t_u\right)}(t_{u+1} - t_u), \quad t_u \in [0, T]$$

For a real signal $F(t) = f(t)$ it is possible to determine the obtained discrete values of time so that they satisfy the following condition and we can define elementary intervals of integration as follows:

$$\int_{t_u(\omega)}^{t_{u+1}(\omega)} f(t)dt = {}^{T}\lambda. \qquad (3)$$

Obviously, the sequences of discrete values of time are completely determined by the function $f(t)$, since they are points at which the integral function $f(t)$ will change on ${}^{T}\lambda(\omega)$. Quite naturally, the number of integration intervals, where corresponding changes have values ${}^{T}\lambda(\omega)$, depends on the ${}^{T}\lambda(\omega)$, as a parameter, and the function $f(t)$. Therefore it can not be determined beforehand. Then we shall obtain evaluation of (1, 2)

$$\tilde{A}(\omega) = {}^{\mathrm{T}}\lambda * \sum_{u=0}^{N_u^A} \cos \omega t_u, \quad u \in \{0, 1, 2, \ldots, N_u^A(\omega)\} \qquad (4)$$

$$\tilde{B}(\omega) = {}^{T}\lambda * \sum_{u=0}^{N_u^B} \sin \omega t_u, \quad u \in \{0, 1, 2, \ldots, N_u^B(\omega)\} \qquad (5)$$

what is more in calculating the values of (4) and (5) is used the same grid values of the analyzed function. Similarly for evaluation of integral (2), it is possible to receive the dual additive formulas for determination of values of function on the given values of its spectral density [2].

$$\tilde{f}(t) = {}^{\Omega}\lambda * \sum_{v=0}^{N_v^{\Omega}} \cos \omega t_v, \quad v \in \{0, 1, 2, \ldots, N_v^{\Omega}(\omega)\}$$

An error value in the integration in the dual additive method depends on the frequency of base functions. The higher the frequency the greater is an error which we obtain for the spectral density.

3 Computer Simulation

We have developed a computer program to simulate of the additive algorithm and have investigated filtering properties of this method calculating the discrete value of the spectrum of a signal. For performing test of additive algorithm, we had to simulate the work analog to digital converter. As mentioned above, this work is carried out only once, and then the results are used repeatedly to compute all digital values of the spectrum.

```
Input:
 xq     - value of the independent variable in a point q
 dxq    - increment of the independent variable
 lambda - threshold value of the integral
 intl    - the current value of the integral
 intlm   - module current value of the integral
 nr      - the current number of the integration interval.
 Nd      - the number of elementary integration intervals
 N       - the number of pixels per image line.
 image(xq)- the brightness of the original image at
            the point q.
Output:
 tableX[Nd]- a table of values of the independent
              variable.
while xq <= N
  while intlm <= lambda
  intl = intl + image(xq+dxq)*dx_q
  intlm = intl > 0 ? intl: - intl
  xq = xq + dxq
  end while
intl = 0
nr++
tableX[nr] = xq
end while
```

Now consider the application of additive algorithm for to calculate DCT conversion, which is widely used for image compression. It is well known, that a DCT is a Fourier-related transform. In general, there are eight standard DCT variants, of which four are often used. Let us consider development DST-II conversion which the most commonly used in modern technologies of compression of visual and audio images, such as JPEG and MPEG [1]. DCT, as cosine transform, assumes an even extension of the original function. In DCT the N real numbers x0, … , xN − 1 are transformed into the N real numbers X0, … , XN − 1 according to the formulas:

$$G(0) = \sqrt{1/N}\sum_{n=0}^{N-1} f(n)$$

$$G(k) = \sqrt{2/N}\sum_{n=0}^{N-1} f(n)\cos(\pi k(2n+1)/2N), \quad k = 1, 2, \ldots N-1. \tag{6}$$

Thus DCT is used to convert f(0), … , f(N − 1) real numbers in a series of real numbers G(0), … , G(N − 1), where G(0) is the average value of the sample sequence.

Now we apply of the formula of the DCT transformation that are described above. According to (4) transform the formula (6) as follows

$$G(k) = \sqrt{2/N}^{T} \lambda \sum_{n=0}^{N-1} \cos(\pi k(2n+1)/2N), \quad k = 1, 2, \ldots N-1. \tag{7}$$

Suppose that at our disposal, there is only a table of values of the original function, which is represented with a uniform step in the independent variable. For testing of the additive algorithm as original function was selected the result of the scan real image. For example, take the result of scanning of a monochromatic color component of the image represented in the system RGB at 24 bits. In this case, each color is represented by 8 bits, therefore brightness of the image points lie in the range 0–255. To estimate of the result of modeling we will compare it with the result of the transformation using DCT algorithm.

In accordance with the idea of additive algorithm a grid of values of the independent variable, in nodes of which should be chosen values of the basic functions is irregular and is determined in the process of integration of the original function. For the simulation of analog digital convertor in this case, we have to perform a linear interpolation in the intervals between the values of the original function, defined in the table of its values. The brightness of the image between neighbors pixels, we obtain using linear interpolation. The numerical value of the integral, we obtain using the method of rectangles. After interpolation, it is necessary to carry out the integration, the resulting function so as to obtain a table of values of the independent variable (tableX[Nd]), just as was done earlier in the program for simulation of analog digital convertor. The text of the main part of the program for modeling of the work of additive algorithm at calculation ADCT transformation for one point of spectrum is represented below.

```
Input:
lambda    - threshold of the integral value
Nd - the number of elementary integration intervals
nr - the current number of interval of integration
k  - the current number of values of the ADCT
tableX[Nd]- a table of values of the independent
               variable.
Output:
ADCT(k)  - digital value of the spectrum in the point k

for(nr = 0; nr <= Nd; nr++)
ADCT(k) = ADCT(k)+
  cos(M_PI*k*(2*tableX[nr]+1)/(2N))
end for
ADCT(k) = ADCT(k)*lambda*sqrt(2/N)
```

In accordance with the above fragment of the program experiment have been performed.

The experimental conditions were as follows: `N ∈ [0, 199]`, `tableX[Nd]`, was defined with parameters `dx = 0.1` and different value of Nd and `lambda`. The simulation results are presented in Fig. 1, on which is shown the graph of the ADCT and graphs of errors, that are displaced downwards along the ordinate on 100 and 200 positions respectively. On graph of errors, displaced downwards along the ordinate on 200 positions down, is represented the experiment, when `lambda` is equal 132. The number of the points, in which we are needed to take the value of the basic function, equal 244. That is in the last fragment of the program code Nd = 244. On graph of errors, displaced downwards along the ordinate on 100 positions down, is represented the experiment, when `lambda` is equal 73. The number of the points, in which we are needed to take the value of the basic function, equal 438. That is in the last fragment of the program code Nd = 438.

In addition in the experiment was conducted calculating of the average error according to formula

$$errorA = (1/N)\sum_{n=0}^{N-1}[DCT(n) - ADCT(n)] \quad (8)$$

and is represented average error modulo according to formula

$$errorAm = (1/N)\sum_{n=0}^{N-1}[DCT(n) - ADCT(n)] \quad (9)$$

The results of the experiment are shown in Table 1. Table provides information as about average error as and about average error obtained by addition modules of the errors on the analysis interval. It is interesting to note that the average error in a certain range is dependent on the number of samples of basic functions Nd. On the other hand average modules errors increases rapidly when the number of samples approaches N, to number of samples of the original signal in DCT algorithm.

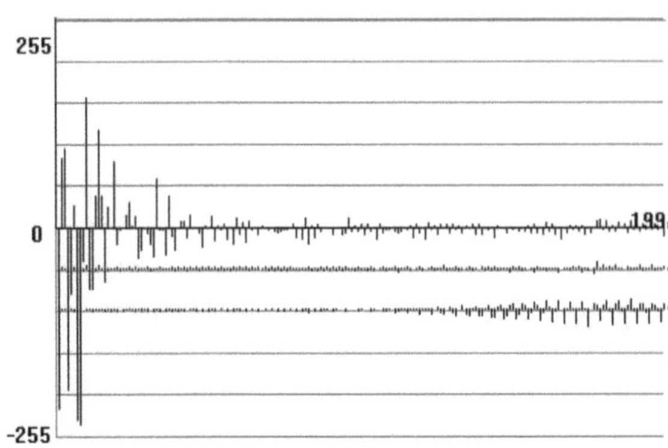

Fig. 1 Spectrum of the original signal, performed by additive algorithm, and graphs of the errors

Table 1 Comparison errors of the additive algorithm

lambda	Nd	errorA	errorAm
64	499	0.11	5.04
73	438	0.35	6.21
96	334	−1.84	5.90
132	244	4.13	9.21

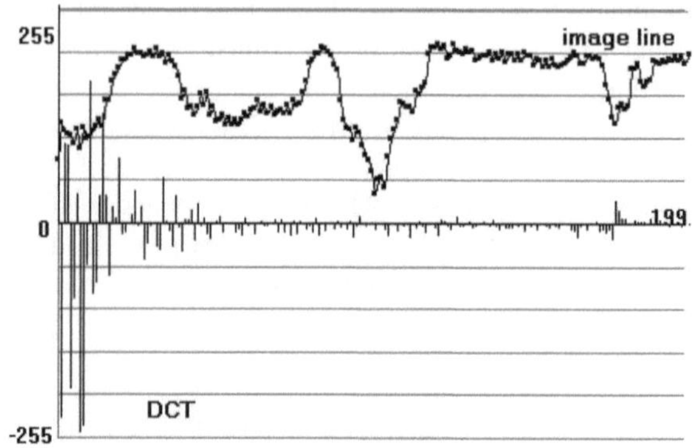

Fig. 2 Line image and DCT spectrum

Now consider the process of restoring an image on the basis of his conversion ADCT. At first, we should to perform the recovery of a one-dimensional function, as which we have been already used one of the horizontal lines of the two-dimensional image. This function is shown in Fig. 2. The discrete spectrum DCT transformation of this function also shown in this figure. By ADCT spectrum we will restore the value of the original function. To do this, we apply of inverse formulas of the DCT transform. The main part of the algorithm for recovery original function is shown below.

```
Input:
N - number of values of the spectrum calculated by ADCT
k - the current number of values of the recovered
    image
n - the current number of values of the ADCT
ADCT(n) - digital value of the spectrum in the point k
Output:
imageR(k)- the brightness of the pixel to the point k
for(n = 0; n <= N-1; n++)
imageR(k) = imageR(k) +
ADCT(n) * cos(M_PI*k*(2*n+1)/(2*N))
end for
imageR(k) = imageR(k)*sqrt(2/N)
```

In accordance with the above fragment of the program, a series of experiments have been performed. The main objectives of the experiment was to get the value of the original function by means of the discrete spectrum of ADCT. After the experiment, we evaluated the average value of the error resulting in the recovery process according to the formulas

$$errorR = (1/N)\sum_{n=0}^{N-1}[\text{image}(n) - imageR(n)] \quad (10)$$

$$errorRm = (1/N)\sum_{n=0}^{N-1}[DCT(n) - ADCT(n)] \quad (11)$$

In Table 2 is represented the errors of the recovered image line, that have been received with using of discrete values of the spectrum, calculated by means of additive algorithm.

The experiments produced the following results.

We also tested the additive algorithm using it for compression of two-dimensional images. In this case, we used the additive algorithm ADCT for image processing in one direction and the conventional discrete algorithm DCT in a different direction. To perform the first step, we had to get a table of values of the independent variable in which the integral of the brightness of the original image is equal to the threshold value. As we analyzed the two-dimensional image it was necessary to obtain a two-dimensional table of values of the independent variable by the direction in which was used algorithm ADCT. To produce a two-dimensional table of values of the independent variable is necessary to use an algorithm similar to that given in the beginning of the article, where it was used to produce a table of values of the independent variable in the one-dimensional table (tableX[nr]). This sequence of values of the independent variable used in the calculation of the discrete spectrum in accordance with the additive algorithm ADCT. The main part of the algorithm is represented below.

Table 2 Comparison of error of the reconstructed image

lambda	Nd	errorR	errorRm
64	499	0.49	3.07
73	438	0.49	3.07
96	334	0.48	4.63
132	244	0.50	5.80

```
Input:
N  - number of point of the image and spectrum in
     direction X(horizontal line)
M  - number of point of the image and spectrum in
     direction Y(vertical line)
Nd -number of intervals of integration in direction X
nd -  the current number of interval of integration
n  -  the current number of values of the image in
     direction X
m -  the current number of values of the image in
     direction
l -  the current number of values of the spectrum in
     direction X
k -  the current number of values of the spectrum in
     direction Y
lambda - threshold of the integral value
table2X[M][Nd] - table of the sequence of discrete values
           of the independent variable in direction N
Output:
table2ADCT[M][N] -table of the value of the spectrum

for(k = 0; k <= M-1; k++)
for(l = 0; l <= N-1; l++)
for(nd = 0; nd <= Nd; nd++)
ADCT2Da=ADCT2Da+
      +lambda*cos(M_PI*l*(2*table2X[k][nd]+1)/(2*N))
end for
table2Da[k][l]=ADCT2Da
for(m = 0; m <= M-1; m++)
ADCT2D= ADCT2D +
       +table2Da[m][l]*cos(M_PI*k*(2*m+1)/(2*M))
end for
table2ADCT[k][l]= ADCT2D
end for
end for
```

The result we received a two-dimensional table of the discrete spectrum.

Further, applying the inverse transformation DCT we perform the restoration the image from its discrete spectrum obtained using additive formulas ADCT. Figure 3 shows the original image and Fig. 4 shows mage, restored from discrete spectrum with using additive algorithm. Monochromatic part of the image obtained on the basis of the green component of the color image.

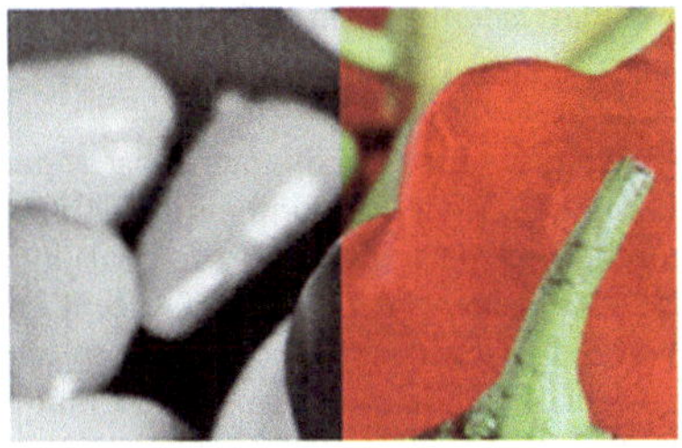

Fig. 3 Original image

Fig. 4 Image, restored from discrete spectrum with using additive algorithm

4 Conclusion

The positive feature of the proposed method is that it can be applied in every integral transform of the signal with different basic functions. For this the original signal have to be converted into a sequence of values of the independent variable, in every of where value of the definite integral from original function are equal to the value of the threshold as described above. Application of the proposed algorithms in practice makes sense if is opportunity to have the values of basic functions in points irregular grid values of the independent variable studied function.

References

1. Shao, X., Johnson, S.G.: Type-IV DCT, DST, and MDCT algorithms with reduced numbers of arithmetic operations. Signal Process. **88**(6), 1553–1564 (2008)
2. Johnson, S.G., Frigo, M.: A modified split-radix FFT with fewer arithmetic operations. IEEE Trans. Signal Process. **55**(1), 111–119 (2007)
3. Puschel, M., Moura, J.M.F.: The algebraic approach to the discrete cosine and sine transforms and their fast algorithms, SIAMJ. Computing **32**(5), 1280–1316 (2003)
4. Guo, Z., Shi, B., Wang, N.: Two new algorithms based on product system for discrete cosine transform. Signal Process. **81**, 1899–1908 (2001)

5. Katkow, A.F.: Dual formulas additive fourier transformation. In: Puchov, G.E. (ed.) Mathematical Modeling and Theory of Electrical Circuit, Naukova Dumka, Kiev, No. 16, pp. 53–59 (1978)
6. Katkov, A., Wegrzyn-Skrzypczak, E.: Speedy numerical algorithms and architecture for additive spectral analysis. J. Math. Model. Algorithms **1**(3), 225–241 (2002)
7. Katkow, A.: Additive algorithms of spectral analysis of signals in real time. In: An Introductory Guide to Image and Video Processing, pp. 251–269. iConcept Press Ltd., USA (2013)
8. Katkow, A.: Additive algorithms for accelerated compression of information. In: Information Systems Architecture and Technology. Selected Aspects of Communication and Computational Systems, pp. 147–158. Oficyna Wydawnicza Politechniki Wrocławskiej, Wrocław 2014 (ISBN 978-83-7493-856-3)

Swarm of Mobile Virtualization Containers

Marian Rusek, Grzegorz Dwornicki and Arkadiusz Orłowski

Abstract A decentralized algorithm for controlling the distribution of tasks between nodes of a simple computational cloud is presented. Each task is running inside a virtualization container and thus can be easily migrated to other hosts. An additional relatively unintelligent mobile agent process is placed inside each container to control this migration process. These mobile virtualization containers use only local interactions and no communication to give the desired global behavior of the cloud providing dynamic load balancing between the servers. The resulting swarm-like group behavior is robust to node failures and to changes in number of tasks running in a cloud.

Keywords Cloud computing · Swarm intelligence · Virtualization containers · Mobile agents

1 Introduction

There is no single definition for what a computer cloud is. For the purpose of this paper we will come up with a very simple working definition which is not meant to be perfect. It says, that a cloud consists of a lot of storage resources with compute cycles located nearby. Essentially, there is so much data that it can not be moved around the internal network of the cloud all that easily. Therefore there is a need to bring the compute cycles closer to the data rather than bring the data closer to where the compute power is.

The idea of moving compute cycles rather than data brings us to the old concept of mobile agents [8]. Mobile agents are programs that can migrate from host to host in a network. The state of a running program is saved, transported to the new server, and restored, allowing the program to continue execution where it left off. Mobile

M. Rusek (✉) · G. Dwornicki · A. Orłowski
Faculty of Applied Informatics and Mathematics, Warsaw University of Life Sciences, Nowoursynowska 166, 02–787 Warsaw, Poland
e-mail: marian_rusek@sggw.pl

© Springer International Publishing Switzerland 2016

J. Świątek et al. (eds.), *Information Systems Architecture and Technology: Proceedings of 36th International Conference on Information Systems Architecture and Technology – ISAT 2015 – Part III*,
Advances in Intelligent Systems and Computing 431, DOI 10.1007/978-3-319-28564-1_7

agent systems differ in a significant way from conventional process migration systems. The agents move when they choose and to the server of their own choice. On the other hand in a typical process migration system the scheduling system decides when and where to move the running process, for example to balance server load in a cluster.

Emerging technologies are making it possible to cheaply manufacture small robots with sensors, actuators and computation. Swarm approaches to robotics, involving large numbers of simple robots rather than a small number of sophisticated robots, has many advantages with respect to robustness and efficiency. Such systems can typically absorb many types of failures and unplanned behavior at the individual agent level, without sacrificing task completion [1–4, 10, 11]. These properties make swarm intelligence an attractive solution also for other problem domains. In this paper we use this approach for task scheduling in a complex distributed system—the cloud.

The classification of the cloud services that are provided is often based on the nature of the services themselves. It is called an "as a Service" (aaS) classification. Infrastructure as a Service (IaaS) allows the user to get access to machines and install his own operating systems inside virtual machines instances without having the security risks of the hardware as a service cloud [14]. Amazon Web Services and Microsoft Azure are popular players that offer services in the infrastructure as a service paradigm. An operating system running inside a virtual machine is completely isolated from the operating system of the host server. It is therefore cumbersome for it to act as a mobile agent, which should be able to initiate its migration process [6, 15].

Recently the industry is moving beyond self-contained, isolated, and monolithic virtual machine images in favor of container-type virtualization [9, 13]. Containers do not contain an operating system kernel, which makes them faster and more agile than virtual machines. Container-type virtualization is an ability to run multiple isolated sets of processes, each set for each application, under a single kernel instance. Having such an isolation opens the possibility to save the complete (in other words, to checkpoint) state of a container and later to restart it. This feature allows one to checkpoint the state of a running container and restart it later on the same or a different host, in a way transparent for running applications and network connections.

In this paper container-type virtualization is used to build a swarm of tasks in a cloud. Each container in addition to the application and their libraries contains an separate process representing the mobile agent. It deals with sensing the neighboring containers and initiating live migration of its container to another host. Containers can call system functions of the operating system kernel running on the server. Therefore in principle they can initiate container migration without help of a separate daemon process running on the host server.

This paper is organized as follows: in Sect. 2 the internals of a virtualization container are analyzed. In Sect. 3 the swarm-like algorithm of container migration in the cloud is introduced. The number of containers on each host plays a role analogous to a pheromone in colonies of insects or simple transceivers mounted on autonomous robots. In Sect. 4 some preliminary experimental results for a simple cloud consisting of two hosts are presented. We finish with a summary and brief remarks in Sect. 5.

2 Container Internals

Virtualization containers are a relatively new technology ideally suited for Platform as a Service (PaaS) and Software as a Service (SaaS) applications [9, 13]. They enable near instant provisioning and on-the-fly modification while delivering near-native application performance, i.e., the virtualization overhead of an application running inside a container is negligible [5]. Virtualization containers provide 10 times the task density per node as compared to virtual machines. Parallels Containers is a widely deployed container-based virtualization software for Linux and Windows operating systems. The results presented in this paper were obtained using an open source version of this software called OpenVZ [9]. OpenVZ is available for Linux operating system only.

There is an alternative container technology on Linux operating system called LXC. Docker, which is based on LXC, was created for different purpose than live migration of containers. Using Docker we can provide an automated application deployment but not live migration.

From a technical point of view containers are basically vehicles for processes. They can contain several processes organized in a hierarchical structure. In this paper we envision a situation when one of these processes contains a mobile agent intelligence and is responsible for sensing other containers and initiating a migration process of the whole container to another node of the cloud.

A virtualization container does not include an operating system kernel—all containers share the kernel of the host (this kernel is running directly on the physical hardware of the server). Processes running inside a virtualization container do not see neither processes running in other containers nor processes running directly in an operating system of the host server. This isolation is achieved using a technology called process namespaces (the concept of namespaces was introduced originally in the Plan 9 operating system [12]). Processes in a Linux operating system are identified by Process Identifier (PID) numbers. Process namespaces in Linux are based on PID isolation. Process cannot access resources shared by processes outside of his namespace because it does not see the PID numbers of these processes. Thus this technique allows to isolate some processes in a way that they will not see other processes running on the system but only the ones inside its process namespace. During migration of a container to another server the PID numbers of the processes running inside it do not change. Therefore a migration process can lead to a situation when we have a Linux system with at least two processes with the same PID numbers. However because of their isolation inside the container's process namespaces no conflict occurs.

Processes running inside the container have their own disk with partitions and file systems. In reality this is a virtual disk and its image is stored as a file on the physical disk of the host. This solution makes the migration of the container's data to another host very easy—only a single file needs to be copied between servers. The network of the container is isolated in a way that allows the container to have they own IP address on the network. This is not the IP address of host but it can be reached from the other containers and hosts. Each container maintains its own state: network connections, file descriptors, memory usage etc. Containers share only the kernel with the host operating system. Thanks to state isolation from other containers a container can be migrated to another system and resumed.

There have been several studies on various optimizations of container migration algorithms [6, 7, 9]. Two best known examples are lazy migration and iterative migration. Lazy migration is the migration of memory after actual migration of container, i.e., memory pages are transferred from the source server to the destination on demand. In the case of an iterative migration of memory happens before actual migration of container. In our experiments with stripped down OpenVZ containers with a size of 50 MB in a test system consisting of two nodes we measured the migration time seen by the host as $T = 6.61$ s and the migration time seen by the container as $\tau = 2.25$ s. The later is three times smaller than the former due to optimizations described above.

When we launch our container for the first time it does not know on what host it was started. However it can use an ICMP echo/reply mechanism to detect the IP address of the host. Each ICMP packet has a TTL (Time-to-Live) value. When this packet is routed trough router this value is decreased. When it reaches 0 the packet is destroyed and an error ICMP packet is send back. This ICMP error packet will have last router IP address. Thus to detect the IP address of the host our container can send an ICMP echo packet with value 1 of TTL to some arbitrary external IP address. The host system acts as a router for container's network. When this special packet is sent by the container it will never reach the destination but the host system will send back an ICMP error packet with its IP address.

Host system keeps all the containers filesystems mounted on its local file system. Each container's file system is visible as a folder which can be found at /var/lib/vz/root/CID. Here CID is a unique container identification number. The location can be exported trough a network file system like NFS. Our container will mount in locally. To do this it needs to know the export path /var/lib/vz/root and the IP address of its host system. By counting the number of entries in this folder it can detect other containers running on the same hosts and count their number N.

3 Swarm Algorithm

In this section we propose a swarm-like algorithm for containers migration. The proposed approach threats the containers as mobile agents and is similar in spirit to the Contained Gas Model describing the motion of pheromone robots [10, 11].

The Contained Gas Model causes agents to disperse evenly within a given shape, and varying the number of agents simply changes the equilibrium density. This model is also capable of automatic self-repair; the system can quickly recover from most patterns of agent death and can receive an influx of new agents at any location without blocking problems.

In our case the shape to be filled by the agents is the entire cloud. Due to the lack of underlying Cartesian space the rules for robot's interactions are replaced by random jumps of the containers between the nodes of the cloud. Each container knows how many hosts (servers) S we have and how many other containers (processes) P are there. These numbers can be updated dynamically at runtime by probing other containers and hosts using ICMP echo/reply protocol. This protocol is implemented by a well known program called ping and thus its messages are often called "pings".

The uniform distribution of containers between hosts gives the following equation for the number of containers n per host:

$$n = \frac{P}{S} \tag{1}$$

Using the method described in Sect. 2, a container can calculate the number N of neighboring containers located on the same host. If $N > n$ the container will try to migrate on another host. The probability of this migration is given by the equation:

$$p = \frac{N - n}{N} \tag{2}$$

As an example let us consider a simple cloud consisting of two hosts. Initially all the containers are launched on the first host. There are no containers on the second host. Equation (2) gives $p = 1/2$ and therefore half of the containers will migrate to the second host in the fist step resulting in their equal redistribution among the hosts.

The complete algorithm executed by a dedicated process running inside the container reads as follows:

1. Use ping to scan the entire subnetwork of the hosts and obtain the number S.
2. Use ping to scan the entire subnetwork of the containers and obtain the number P.
3. Calculate desired container density n according to Eq. (1).
4. Use the method described in Sect. 2 to get the IP address of the host.
5. Mount the /var/lib/vz/root folder exported by the host.
6. Obtain the number of neighboring containers N by counting the number of entries in this folder.
7. If $N \leq n$ go to 15.
8. Calculate the probability of migration p according to Eq. (2).
9. Generate a random number $0 < r < 1$.

10. If $r<p$ then ask the host to migrate to a randomly chosen host (excluding the current one). Otherwise go to 15.
11. Get the IP address of the host.
12. If it's different from the previous one then the migration to another host is complete—go to 13. Otherwise wait t'' seconds and go to 11.
13. Unmount /var/lib/vz/root from the old host.
14. Mount it from the new host and go to 6.
15. Wait t' seconds and go to 6.

In the standard OpenVZ distribution the migration process of a container is initiated by a shell command running outside of it. Thus a container can not initiate the migration process. To make this possible we have added a special system function to the hosts operating system kernel. In the algorithm above we assume, that this call (performed in step 10) is a nonblocking one.

4 Experimental Results

The experiments were performed on $S = 2$ servers equipped with Intel Core i5-2400 CPU, 2 GB of RAM each connected by a dedicated Gigabit Ethernet network. Both servers were running Debian GNU/Linux operating system with OpenVZ software installed. The Linux kernel was modified by adding some new system functions including one described in Sect. 3. At the initial time $P = 16$ identical containers are launched on the first host. Each container has a size of about 1 GB and we measured that its migration to the second host takes $T = 32.5$ s. Starting all the containers takes about 1 min or 2T. After this time on each container a Python script implementing the algorithm from Sect. 3 is running. It starts by scanning the network using nmap to find the number of containers P, as well as the number of hosts S and their IP addresses. Then the mean number of containers n is calculated and the script enters a loop in which it periodically checks the number of neighboring containers N, and decides whether to migrate to the second host or not. In addition, on the host server a monitor program was started which periodically checked the number of containers N in the filesystem and the number of containers queued Q for migration in the kernel queue (access to this data from a user process was possible by a custom system function added to the kernel). From now on the time t' is equal to 2 s.

In Fig. 1 we have the number of containers N on the first host plotted versus time t. In addition, the number of containers Q is also plotted. It is seen from inspection of this plot that for $0<t<2T$ during the startup process the number of containers is constant $N = 16$. This can be explained by the fact that the containers images were initially present on the filesystem but not active. Next, a migration process starts—the steps visible on the plot correspond to the change of N by 1 and change of t by approximately T. Notice that after reaching an equilibrium number $N = P/2$ at $t \simeq 12$T the migration still continues. Four excess containers are migrated to the

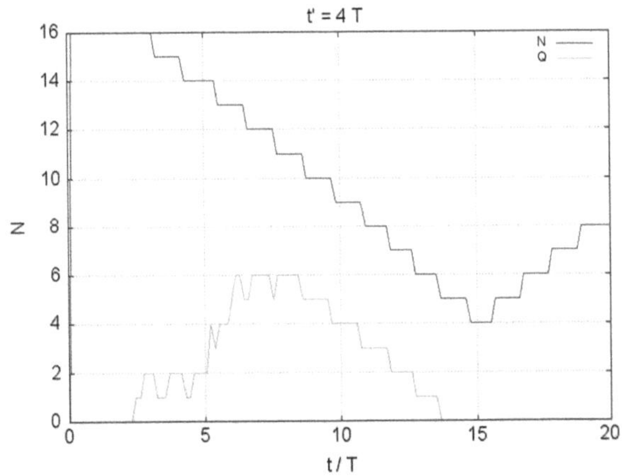

Fig. 1 Number of containers on the first host versus time

second host and then some of them come back. Equilibrium is reached only at $t_0 \simeq 18\mathrm{T}$. Thus it takes more than twice as long as migration of the $n = 8$ containers in a loop.

We performed several such experiments for different sleeping times t'. In Fig. 2 we plot time t_0 needed to reach equilibrium as a function of the sleeping time t'. We see that the algorithm stabilizes only for $t' > 2T$. Result presented in Fig. 1 corresponds to $t' = 4\mathrm{T}$. There are also large fluctuations in the values of t_0—they turn out very sensitive to the initial conditions. For different runs of the experiments at the same values of t' we get different values of t_0. This unstable behavior of the algorithm can be understood by an observation that during migration the container's files are present on both hosts. Therefore the total number of containers seen by the migration agents running inside the containers is not conserved.

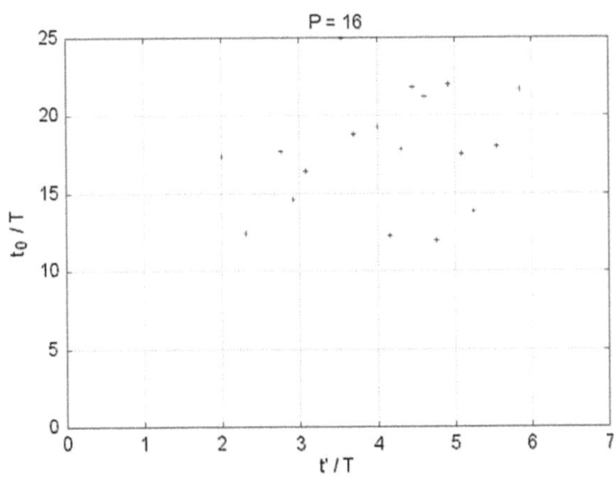

Fig. 2 Time needed to reach equilibrium versus sleeping time

To deal with this problem a modification of the algorithm is needed. To prevent the containers from being counted twice, what affects Eq. (2), we subtracted from the number of containers N present in the filesystem the number of containers Q queued for migration. In this way the migration process of the containers appears to be instantaneous to its neighbors. The new migration probability reads as

$$p' = \frac{N - Q - n}{N - Q} \tag{3}$$

The results of experiments using this modified algorithm are presented in Figs. 3 and 4. Equilibrium in Fig. 3 is reached almost as fast as in a centralized algorithm based on a sequential migration in a loop. As seen in Fig. 4 algorithm is now very robust and the time needed to reach equilibrium $t_0 \simeq 10\mathrm{T}$ is almost independent from the sleeping time t'. However there is still some instability. It becomes especially pronounced for $t' > 5T$.

To understand its origin let us investigate in detail the case corresponding to the first instability point from Fig. 4. It happened at $t' \simeq 0.61T$ and the time needed to reach equilibrium $t_0 \simeq 18T$ was almost twice as large as the usual value of $10T$. In Fig. 5 we plot the time dependence of number of container's images N_i and number of queued containers Q_i on both hosts ($i = 1, 2$). After reaching equilibrium value of $N_1 = N_2 = 8$ two additional migrations from the first host to the second and back occurred. It turns out that this is due to an error in determining the total number of containers P by one of the mobile agents. It was started at the end scanned the network while one of the first containers was already migrating. At some period of time $\tau < T$ during migration the container is invisible from the network. Thus the mobile agent obtained the number $S = 15$ and the equilibrium density $n = 7$. This caused it to jump back and forth between the servers with a small probability $p = 1/8$.

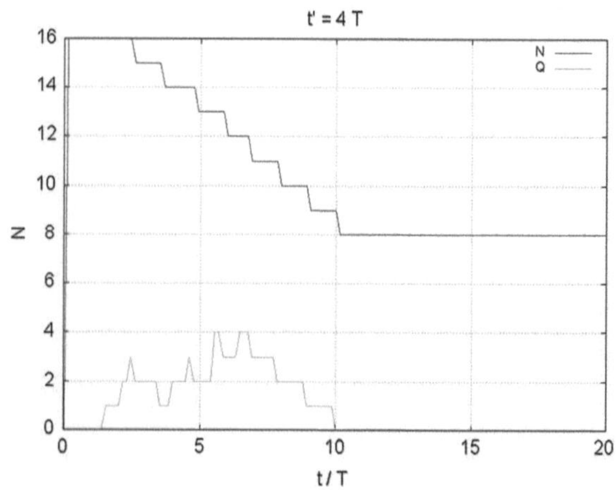

Fig. 3 Number of containers on the first host versus time. Modified algorithm

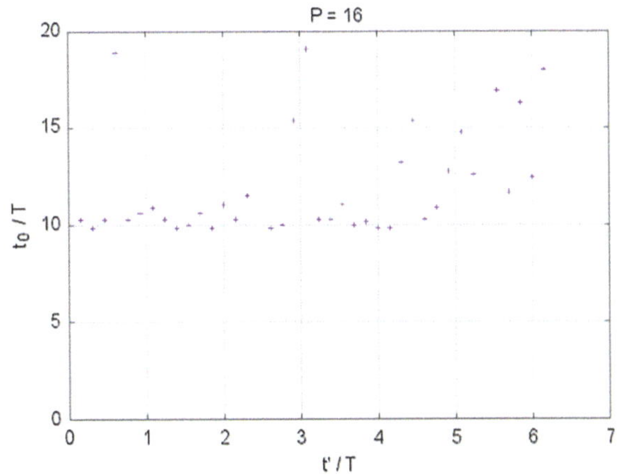

Fig. 4 Time needed to reach equilibrium versus sleeping time. Modified algorithm

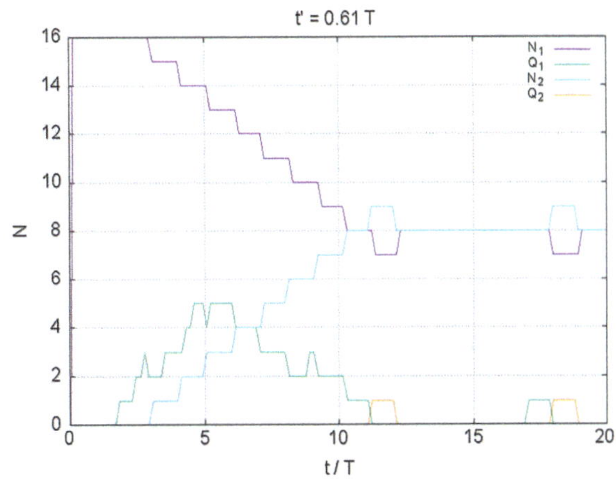

Fig. 5 Number of containers on both hosts versus time. Modified algorithm

5 Conclusion

In summary, the OpenVZ containerization software was used to implement a swarm of tasks executing in a simple cloud. Each task includes a mobile agent process which governs its migration to the other node of the cloud. The tasks running on the nodes of the cloud self-organize to maintain a constant load among the servers. The system automatically adapts to creation and destruction of tasks as well as extension of the cloud by new servers. It can be easily adopted to react on server failures: a failing server can produce an artificial pheromone by creating entries in the / var/lib/vz/root directory of its filesystem. This will cause all the tasks running on it to migrate away from the pheromone.

The performance of the swarm-like algorithm proposed to control the containers was experimentally tested on a simple "cloud" consisting of two nodes. Initially all the containers reside on the first node—the time needed to reach a stationary state of equal redistribution of the containers between the nodes is measured. The obtained results are compared to a centralized algorithm implemented as a shell script. This script monitors the contents of /var/lib/vz/root directories on both hosts and in the case of non equal numbers of entries initiates migration of excess containers.

In future work, we would like to generate a consensus map of the cloud through agents performing local measurements of the speed of network links to other nodes and counts of the number of neighboring tasks. Such a map will have a form of a weighted graph with vertices labeled by the numbers of tasks on each node and edges labeled by the speeds of network links connecting the nodes. Such a map will allow the containers to migrate in a more intelligent way and thus reduce the time needed by the system to reach equilibrium after creation or termination of tasks and additions or failures of the nodes.

References

1. Berman, S., Halász, A., Kumar, V., Pratt, S.: Bio-inspired group behaviors for the deployment of a swarm of robots to multiple destinations. In: 2007 IEEE International Conference on Robotics and Automation. pp. 2318–2323. IEEE (2007)
2. Cheah, C.C., Hou, S.P., Slotine, J.J.E.: Region-based shape control for a swarm of robots. Automatica **45**(10), 2406–2411 (2009)
3. Cheng, J., Cheng, W., Nagpal, R.: Robust and self-repairing formation control for swarms of mobile agents. AAAI **5**, 59–64 (2005)
4. Dorigo, M., Trianni, V., Sahin, E., Groß, R., Labella, T.H., Baldassarre, G., Nolfi, S., Deneubourg, J.L., Mondada, F., Floreano, D., et al.: Evolving self-organizing behaviors for a swarm-bot. Auton. Robots **17**(2–3), 223–245 (2004)
5. Hacker, T.J., Romero, F., Nielsen, J.J.: Secure live migration of parallel applications using container-based virtual machines. Int. J. Space-Based Situat. Comput. **2**(1), 45–57 (2012)
6. Hines, M.R., Gopalan, K.: Post-copy based live virtual machine migration using adaptive pre-paging and dynamic self-ballooning. In: Proceedings of the 2009 ACM SIGPLAN/SIGOPS International Conference on Virtual Execution Environments, pp. 51–60. ACM (2009)
7. Hirofuchi, T., Nakada, H., Itoh, S., Sekiguchi, S.: Reactive cloud: Consolidating virtual machines with postcopy live migration. Inf. Media Technol. **7**(2), 614–626 (2012)
8. Kotz, D., Gray, R.S.: Mobile agents and the future of the internet. Operat. Syst. Rev. **33**(3), 7–13 (1999)
9. Mirkin, A., Kuznetsov, A., Kolyshkin, K.: Containers checkpointing and live migration. In: Proceedings of the Linux Symposium, pp. 85–92 (2008)
10. Payton, D., Estkowski, R., Howard, M.: Progress in pheromone robotics. Intell. Auton. Syst. **7**, 256–264 (2002)
11. Payton, D., Estkowski, R., Howard, M.: Compound behaviors in pheromone robotics. Robot. Auton. Syst. **44**(3), 229–240 (2003)
12. Pike, R., Presotto, D., Thompson, K., Trickey, H., Winterbottom, P.: The use of name spaces in plan 9. In: Proceedings of the 5th Workshop on ACM SIGOPS European Workshop: Models and Paradigms for Distributed Systems Structuring, pp. 1–5. ACM (1992)
13. Scheepers, M.J.: Virtualization and Containerization of Application Infrastructure: A Comparison (2014)

14. Zhang, Q., Cheng, L., Boutaba, R.: Cloud computing: state-of-the-art and research challenges. J. Internet Serv. Appl. **1**(1), 7–18 (2010)
15. Zhao, M., Figueiredo, R.J.: Experimental study of virtual machine migration in support of reservation of cluster resources. In: Proceedings of the 2nd International Workshop on Virtualization Technology in Distributed Computing. p. 5. ACM (2007)

Towards Healthcare Cloud Computing

Piotr Dorożyński, Adam Brzeski, Jan Cychnerski and Tomasz Dziubich

Abstract In this paper we present construction of a software platform for supporting medical research teams, in the area of impedance cardiography, called IPMed. Using the platform, research tasks will be performed by the teams through computer-supported cooperative work. The platform enables secure medical data storing, access to the data for research group members, cooperative analysis of medical data and provide analysis supporting tools basing on intelligent information retrieval techniques (semantic text comparison, charts, images, video matching). On top of the platform, a dedicated web application is built for medical teams focused on the area of hemodynamics. The acquired data enables the researchers to develop new classification schemes for hemodynamic profiles, which will be used for construction of prototype application for supporting stroke treatment centers. In the paper we compare IPMed to other existing platforms which offer similar functionality and describe a concept of its migration to a cloud based system.

Keywords Service oriented architecture · Medical systems · Cloud computing

1 Introduction

Actual state of medicine, diagnostics and medical imaging leads to abrupt growth of diagnostic data available to physician or medical research teams. The amount of this data is often so big that in many cases physician and researchers are not able to

P. Dorożyński · A. Brzeski · J. Cychnerski · T. Dziubich (✉)
Department of Computer Architecture, Gdańsk University of Technology, Gdańsk, Poland
e-mail: dziubich@eti.pg.gda.pl

P. Dorożyński
e-mail: piotr.doro.dorozynski@gmail.com

A. Brzeski
e-mail: brzeski@eti.pg.gda.pl

J. Cychnerski
e-mail: jan.cychnerski@eti.pg.gda.pl

© Springer International Publishing Switzerland 2016

J. Świątek et al. (eds.), *Information Systems Architecture and Technology: Proceedings of 36th International Conference on Information Systems Architecture and Technology – ISAT 2015 – Part III*,
Advances in Intelligent Systems and Computing 431, DOI 10.1007/978-3-319-28564-1_8

analyze and interpret it within acceptable accuracy and time. To address this problem, one of emergent trends in medical IT systems is building private cloud systems, which can store, archive and process this data on-the-fly.

Compared to other businesses, the public healthcare service has significantly underused technology to improve operational efficiency. In Poland, there still exist systems which rely on paper medical records, but this situation is rapidly changing. Information that is digitized is typically not portable, inhibiting information sharing amongst the different healthcare actors. Use of technology to facilitate collaboration amongst the medical community is limited. The healthcare industry is shifting toward an information-centric care delivery model, enabled in part by open standards that support cooperation, collaborative workflows and information sharing. Thus, healthcare information systems (HIS) should be modernized and cloud computing is at the center of this transformation [1].

Cloud computing provides an infrastructure that allows hospitals, medical practices and research facilities to tap improved computing resources at lower initial capital outlays. Additionally, cloud environments reduce the barriers for innovation and modernization of HIS systems and applications.

Many university and pharmacology enterprises are starting to use the cloud to improve research and drug development. The growth in medical data from next generation sequencing as well as the growing importance of biologics in the research process is making cloud-based computing an increasingly important aspect of research and development (R&D). Currently, such institutions do not have the capacity to store and process large datasets. Physician collaboration solutions such as remote video conference physician visits are well-known [2–4]. Cloud technology supports medical teams collaboration and gives the ability to use applications based on exchanging or sharing large sets of medical cases. This technology assures the required security and privacy level. Cloud computing facilitates dissemination of information and insights in near real-time. The most current, complete insights and clinical knowledge are available to support care provider decisions and to enable a focus on the real problems not on tools. Information contained within a cloud can be better analyzed by using computationally efficient methods and tools.

2 Existing Solutions

The amount of digital information which is collected and stored in modern hospitals poses a great opportunity for extensive data analysis, enabling medical researchers to discover relationships in the data, extend knowledge of the diseases and therefore improve the treatment. However, as a result of numerous legal and administrative difficulties in collecting medical records, sensitive medical data is rarely gathered by hospitals in a cooperative manner. This also prevents establishment of commercial systems. Though, examples of emerging systems of that kind can be already found in the literature.

Rolim et al. [5] presented a concept of a system for automatic collecting and storing medical data from medical devices by using wireless sensors. Authors propose system architecture with a cloud system for storing the data and enabling its analysis by medical experts, accompanied with local servers located in hospitals for gathering data from sensors and making it accessible to medical staff. The system however relies on the possibility of straightforward connecting the wireless sensors to medical devices, without considering the diversity of the interfaces and potential problems in collecting the data. Pandey et al. [6] proposed a cloud system for collecting and analyzing ECG data from users' wearable devices. The solution serves as a health monitoring system. Data analysis is possible via dedicated services hosted in the cloud system. The authors emphasize the requirement of providing high scalability of the system, which is assured by implementing a cloud computing architecture. Patel et al. [7] presented a system for collecting medical images along with their supplementing data in a centralized database. The system enables efficient storing of the images, reducing the cost of gathering the dataset for medical research. The system is demonstrated on the case of acquisition of digital mammography images.

Another type of systems utilize cloud architecture strictly for providing data analysis services. Hsieh and Hsu [8] proposed a cloud based solution for tele-consulting 12-lead ECG examinations on mobile devices. Shen et al. [9] presented a Service Oriented Architecture (SOA) system for EEG signals classification. Lai et al. [10] used Knowledge as a Service (KaaS) model for providing a collaborative system for facilitating radiotherapy dynamic treatment service. Finally, noteworthy systems have been also proposed strictly for the analysis of large collections of data. In [11] we reviewed existing data mining applications for knowledge extraction from medical data. Also, dynamically developing neural networks and deep learning methods are recently increasingly used for data analysis. We give more attention to this area later in the text.

Unlike the systems listed above, our system is based on web services (software plugins) for professional healthcare devices. Our goal is acquisition of valuable data for medical researchers which are hard to collect. A single doctor has only a small number of valuable research (often tens of). Our platform allows to collect hundreds of examinations. Data obtained from personal medical devices used in the papers mentioned above have also a high degree of uncertainty as compared to the data from the hospitals.

3 The IPMed Platform

The proposed system was designed as a thorough solution for medical researchers seeking to gather and analyze large amounts of diverse medical data. Currently, the system was successfully deployed in three independent hospital wards, where it enables gathering extensive data of stroke cases, including impedance cardiography

(ICG) examination records. The general functionality of the platform can be outlined as follows:

- Medical data gathering. The platform enables flexible extending with dedicated modules for communicating with particular medical devices as well as front-end interfaces for entering data by medical staff.
- Medical data storing. The data is automatically stored on a local server at the hospital as well as sent to a centralized database, ensuring data security and persistence.
- Data analysis supporting tools. The data collected in the centralized database can be analyzed by medical researchers using a set of statistics and data mining algorithms.
- Remote teamwork of researchers. The platform enables organization of medical research groups for cooperative working on the data.

The data analysis and teamwork modules are being still extended in order to possibly well respond to medical researchers' needs. High versatility and flexibility of the platform is worth emphasizing, as it is designed for handling wide range of medical data types and therefore can be applied in various areas of medicine.

3.1 General Architecture

The IPMed system is based on a component structure, in which every component is responsible for a particular task. The components communicate with each other through network using REST (Representational State Transfer) architecture style. The communication is secured by encryption and users are authenticated by login and password pair. The databases that contain sensitive data (e.g. medical data) are encrypted and the access to the key is only granted to authenticated and authorized users, which in turn keep data safe even in case of hardware theft. Figure 1 presents the overview of the main IPMed components and their interconnections. Detailed description of every component is presented in the next section.

3.2 Components

The IPMed system contains several basic component types depending on their role. The central point of the system is the main repository, which gathers anonymous data that it receives from local hospital data servers. Each hospital that participates in the grant gathers data from medical equipment and physicians. The data stored in these hospitals is encrypted and decryption requires an authorized user to authenticate. On the other side of the system, there is a cooperation webserver that serves a website for physicians cooperation.

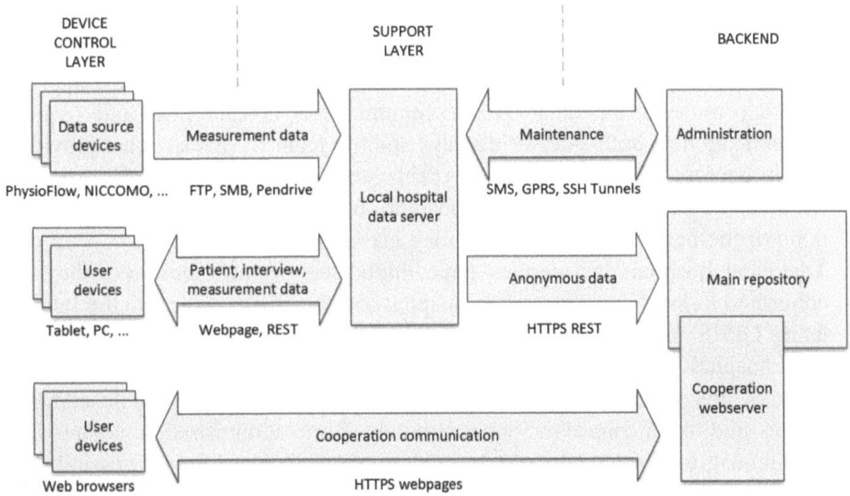

Fig. 1 Main IPMed components and their interconnections

1. **Main repository**. It is a well secured webserver, which is only accessible by applications that possess trusted encryption key that is used for encrypting communication. All the data kept on this server is completely anonymous (anonymization of this data is done on local hospital data servers before this data ever leaves the source hospital). The repository gathers data from hospitals along with its whole history using RESTful web services and the communication is realized using encrypted HTTPS protocol. The gathered data is then directly served to the cooperation webserver. The main repository also provides access to statistical tools that may be used for analysis of gathered data.
2. **Local hospital data servers**. Each hospital has its own local data server, which contains patients' medical data that are stored in an encrypted database. This is the only part of the system that stores sensitive data. The key used for encrypting the database is also encrypted, but it is done using users' keys that are not stored on the same server. This way the access to the database is only possible after authorized user authenticates using his key and thus the data is safe even after hardware theft or physical access to the database files.
   Local hospital data server enables to automatically download and convert data from medical equipment (PhysioFlow and NICCOMO cardiographs) through FTP, SMB or USB flash drive. Downloaded data is temporarily stored in a non encrypted queue (since the key to the database is encrypted it is impossible to open the database until an authorized user authenticates using his key and thus decrypt the database key), but as soon as the database is opened the downloaded data is moved to the database and encrypted. The data stored on a local hospital data server are exposed through RESTful web services that enables their browsing and modification, but requires devices and applications (e.g. dedicated

Android client application) to authenticate using trusted key and requires user (i.e. physician) to authenticate using login and password pair.
Each local hospital data server connects to the main repository in fixed intervals in order to send the data. This communication is encrypted and requires authentication of both sides using keys and the main repository must provide a key to decrypt the local database. At this step all of the data is also carefully anonymized before they leave the hospital, because it is a legal obligation that none of the patient's personal data may leave the hospital.
The local hospital data servers have duplicated network links i.e. they are connected to local area network in hospital and they may connect to the Internet using GPRS modem. This redundancy was primarily done in order to make local hospital data servers as reliable as possible, because they were designed to work as independent devices that no one is working on and whole administration and maintenance is done remotely from administration centre. The requirement for remote administration also created the need for the possibility of some basic administration tasks (getting error logs, restarting the device etc.) even when the device is not connected to the network and so it is done using SMS messages.

3. **Administration centre**. This is the special server that is used for administration purposes. All local hospital data servers connect to this one and creates SSH (Secure Shell) tunnel (using trusted keys) that may then be used for maintenance and administration.
4. **Data source devices**. These are the devices that gather medical data. They are connected to local hospital data servers and upload data to them. In prototype application we use Manatec PhysioFlow and Medis NICCOMO cardiographs as our data source devices.
5. **User devices in hospitals**. Data stored on local hospital data servers are accessed via user devices like tablets with Android or iOS or other devices with web browser (see Fig. 2). The devices have preinstalled IPMed client applications, which connect to the local hospital data server via RESTful web services. Similarly like in the other places in the system, the devices have to authenticate using trusted keys. Apart from device authentication, users also have to authenticate using login and password pair. To make sure that no data will be stolen if the device is stolen, none of the data (medical data, passwords, keys etc.) is stored persistently on the device.
6. **Cooperation web server**. This web server provides a platform for cooperation of physicians from different medical facilities by giving them access to the data stored in the main repository in one common and uniform data format. The platform also contains a recommender system based on expert system that classifies patients by their hemodynamic profile. This part of the system enables research for both physicians and computer scientists. From physicians' point of view the cooperation platform enables to share knowledge and medical data that

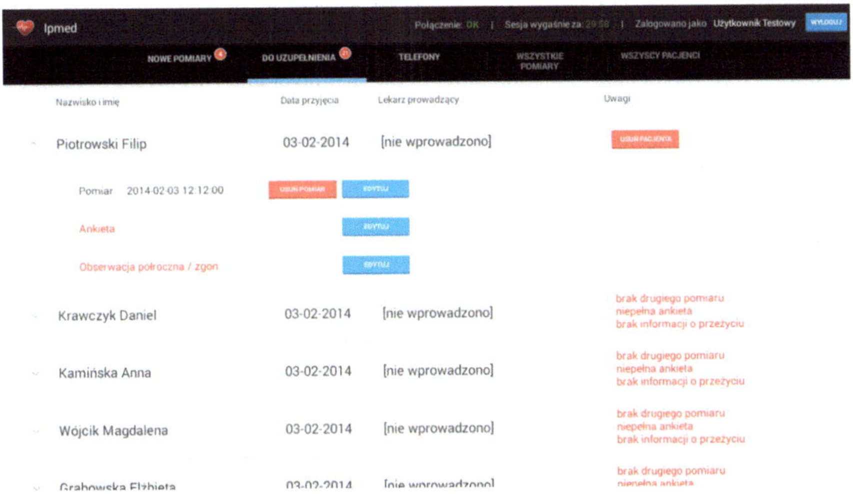

Fig. 2 Screen of the data management frontend in the Android application for stroke patients data gathering

is needed for research. On the other hand, from the computer scientists' point of view it enables to use gathered data in research on artificial intelligence, particularly for finding regularities in the data that may be useful for physicians using this system.

It is planned that in the future this platform will give possibility to query central repository for similar cases to a provided one, e.g. finding patients with probably the same disease based on his measured ECG (electrocardiography) signal, using artificial intelligence, especially deep learning techniques. This feature is currently in a design stage and for our pilot application (electrocardiography) the literature does not provide direct solution to the problem. There is a large number of publications on the topic of biometric recognition of human identity using ECG [12–14], which is a similar task and mostly uses hand designed features of ECG from which most common are temporal features (e.g. length of QRS complex), amplitude features (e.g. amplitude of R peak) and slope features (e.g. slope of R peak). However, finding similar cases requires extracting more high level features as otherwise it will consider two cases to be similar only if they come from the same patient and so it seems more suitable to use machine learning methods instead. There were some attempts for automatic recognition of diseases using machine learning, e.g. Dorffner et al. [15] used recurrent as well as feedforward (with hand designed features) neural networks for recognizing ischemic heart disease and they achieved results even better than skilled cardiologists with slight advantage of recurrent over feedforward network. Silipo et al. [16] also used neural networks for disease recognition. They tried to recognize arrhythmia, ischemia and chronic myocardial diseases and they used 3 different networks for each of the tasks. They used feedforward network with

R-R intervals as input for arrhythmia, recurrent network for detecting ST segment anomalies caused by ischemia and a hybrid feedforward network with normalized radial basis function units in the first layer and two layers of standard sigmoid units above it. However, in our case we will need to use one method that will be able to extract features that in general finds good discrimination between patients with different diseases.

4 The IPMed Platform as a Cloud

Our platform can be divided into three parts: device control layer, support layer and backend (as shown on Fig. 1). In the nearest future we plan to migrate the backend to a private medical cloud platform. Cloud computing delivers infrastructure, platform, and software as service and can be classified into SaaS, PaaS and IaaS. Chen contended that cloud business models include medical cloud, education cloud, telecom cloud, financial cloud, manufacture cloud and logistics cloud [17]. The following six domains for the application of cloud computing to healthcare were identified: telemedicine/teleconsultation, medical imaging, public health and patients' self-management, hospital management/clinical information systems, therapy and secondary use of data. Our kind of cloud is classified as secondary use of data cloud systems. This domain contains cloud computing utilization for enabling secondary use of clinical data, e.g. for data analysis, text mining or clinical research.

Business models include private cloud service, public cloud, common cloud and hybrid cloud. We plan to build a private medical cloud based on our HPC lab with a small cluster consisting of 50 nodes with four Intel Pentium III Xeon processors (700 MHz, 1 MB Cache L2). Nodes are connected using Scalable Coherent Interface. We use Linux Virtual Server as load-balancing software running on Debian Linux, Xen Software for virtualization, OpenNebula as management and monitoring software and GlassFish as web application server are used. We treat the entire cluster as PaaS.

Currently, the gathered data is stored in Postgresql but to achieve better scalability we plan to use the Hadoop platform to solve the exchanging, storing, and sharing issues in medical data (especially images and videos) (see Fig. 3).

By using Mule ESB as the communication method between components/services we can extend the platform with additional components such as modeling analytics tools and high performance algorithms support, for example, deep learning methods which were mentioned in Sect. 3.2. Service oriented architecture enables easy integration with other services e.g. data converters increasing of HIS interoperability or decision support services with a desired level of reliability [18].

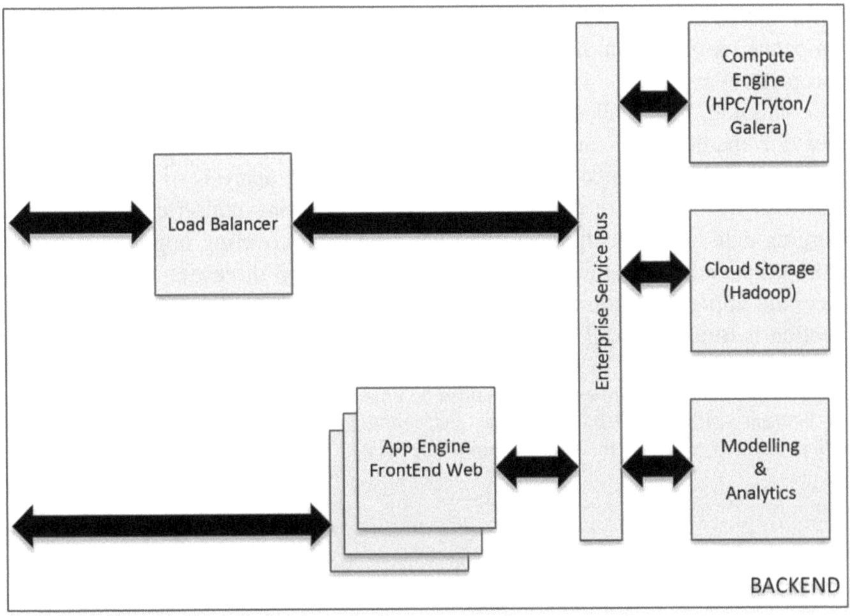

Fig. 3 IPMed planned components in cloud environment

5 Conclusions and Future Work

The IPMed platform incorporates four neurological wards in Polish hospitals, enabling gathering of data of stroke cases, including ICG. Hemodynamic disturbances occurring in 60–100 % of 70,000 patients with stroke are the second cause of death, responsible for 15 % of deaths in hospital phase. Fifty-six complete examinations were collected so far. The results of examination are related to 45 patients with ischaemic stroke (TIA) and 16 patients with hemorrhagic stroke. Statistical significance is not achieved in both groups. Everything indicates that this is due to too small sample size and further confirms the desirability of further research. Detailed results of clinical research are presented in [19].

In this paper we have presented a description of a service oriented medical platform supporting cooperative operation for medical teams, currently focused on the area of stroke treatment, ICG examinations and hemodynamics. We have described the IPMed platform and shown the process of migration to a cloud system. In the near future we plan hardware configuration, low-level software installation (PaaS) and adaptation of the IPMed platform to cloud service (SaaS).

The following concerns in this case of cloud computing application can be identified: safety/security of data as a threat to privacy, reliability and transparency of data handling by third parties. In the design of our system we put a lot of effort to ensure high level of data safety and security. Also, service oriented architecture allows improving system flexibility and its extension at low cost (with third-parties data converters), we can therefore consider the requirements to be addressed in an appropriate way.

Concluding, the development of data gathering platforms opens great opportunities for medical data analysis. Big sets of medical data have a tremendous importance for the science. Not only a regular statistic analysis of the data can be performed by medical researchers, but also a great contribution can be made by emerging data mining techniques. This can lead to discovering important relations in the data and developing new treatments for many diseases. Therefore, after successful implementation of the system in the area of stroke-related data, the future direction is to apply the IPMed platform in other fields of medicine.

Acknowledgements This research was funded by grants from National Centre for Research and Development (PBS2/A3/17/2013, Internet platform for data integration and collaboration of medical research teams for the stroke treatment centers).

References

1. Priyanga, P., Muthu Kumar, V.: Cloud computing for healthcare organization. Int. J. Multi. Res. Dev. **2**(4) Part J, 487–493 (2015)
2. VSee, a telemedicine platform. http://vsee.com
3. Czekierda, Ł., Masternak, T., Zielinski, K.: Evolutionary approach to development of collaborative teleconsultation system for imaging medicine. In: IEEE Transactions on Information Technology in Biomedicine, pp. 550–560 (2012)
4. Taddei, A., Gori, A., Rocca, E., Carducci, T., Piccini, G., Augiero, G., Festa, P., Assanta, N., Ricci, G. Murzi, B.: Telemedicine network for early diagnosis and care of heart malformations. In: The International Conference on Health Informatics, IFMBE Proceedings, vol. 42, pp. 268–271, Springer International Publishing (2014)
5. Rolim, C.O., Koch, F.L., Westphall, C.B., Werner, J., Fracalossi, A., Salvador, G.S.: A cloud computing solution for patient's data collection in health care institutions. In: Second International Conference on eHealth, Telemedicine, and Social Medicine. ETELEMED'10. IEEE (2010)
6. Pandey, S., Voorsluys, W., Niu, S., Khandoker, A., Buyya, R.: An autonomic cloud environment for hosting ECG data analysis services. Future Gener. Comput. Syst. 147–154 (2012)
7. Patel, M.N., Looney, P., Young, K., Halling-Brown, M.D.: Automated collection of medical images for research from heterogeneous systems: trials and tribulations. In: SPIE Medical Imaging. International Society for Optics and Photonics (2014)
8. Hsieh, J.C., Hsu, M.W.: A cloud computing based 12-lead ECG telemedicine service. In: BMC medical informatics and decision making, p. 77 (2012)
9. Shen, C.P., Chen, W.H., Chen, J.M., Hsu, K.P., Lin, J.W., Chiu, M.J., et al.: Bio-signal analysis system design with support vector machine based on cloud computing service architecture. In: Proceedings of the EMBC, pp. 1421–1424 (2010)
10. Lai, I.K., Tam, S.K., Chan, M.F.: Knowledge cloud system for network collaboration: a case study in medical service industry in China. In: Expert Systems with applications, pp. 12205–12212 (2012)
11. Brzeski, A., Cychnerski, J., Rościczewski, P.: Data mining applications and methods in medicine. In: 35th International Conference Information Systems Architecture and Technology (2014)

12. Tantawi, M., Salem, A., Tolba, M.F.: ECG signals analysis for biometric recognition. In: 14th International Conference on Hybrid Intelligent Systems, pp. 169–175 (2014)
13. Biel, L., Pettersson, O., Philipson, L., Wide, P.: ECG analysis: a new approach in human identification. In: IEEE Transactions Instrumentation and Measurement, pp. 808–812 (2001)
14. Israel, S.A., Irvine, J.M., Cheng, A., Wiederhold, M.D., Wiederhold, K.: ECG to identify individuals. Pattern Recogn. **38**(1), 133–142 (2005)
15. Dorffner, G.; Leitgeb, E.; Koller, H.: Toward improving exercise ECG for detecting ischemic heart disease with recurrent and feedforward neural nets. In: Proceedings of the 1994 IEEE Workshop: Neural Networks for Signal Processing IV., pp. 499–508, (1994)
16. Silipo, R., Marchesi, C.: Artificial neural networks for automatic ECG analysis. IEEE Trans. Signal Process. **46**(5), 1417–1425 (1998)
17. Yang, C.T., Chen, L.T., Chou, W.L., Wang, K.C.: Implementation of a medical image file accessing system on cloud computing. In: 13th International Conference on Computational Science and Engineering, pp. 321–326 (2010)
18. Brzeski, A., Cychnerski, J., Dziubich, T., Kaczmarek, P.: Dependable integration of medical image recognition components. In: Świątek, J., Borzemski, L., Grzech, A., Wilimowska, Z.: Information Systems Architecture and Technology: System Analysis Approach to Design, Control and Decision Support. Oficyna Wydawnicza PWr, pp. 151–160, Wrocław (2012)
19. Siebert, J., Molisz, A., Trzeciak, B., Gutknecht, P., Dziubich, T., Krawczyk, H.: Program IPMed—zastosowanie kardiografii impedancyjnej do monitorowania układu krążenia. Forum Medycyny Rodzinnej, pp. 1–12 (2015)

Parallelization of Divide-and-Conquer Applications on Intel Xeon Phi with an OpenMP Based Framework

Paweł Czarnul

Abstract The paper proposes an approach for efficient and flexible parallelization of divide-and-conquer computations using the modern Intel Xeon Phi accelerators. Many real-life problems follow the divide-and-conquer paradigm and consequently generate either balanced or imbalanced trees. The paper proposes an OpenMP multi-threaded implementation of a general framework that requires coding basic divide-and-conquer constructs such as data partitioning, computations and result integration. Mapping computations onto threads is handled by the underlying runtime layer. The paper presents performance results for a parallel adaptive quadrature integration resulting in an irregular and imbalanced tree. It is shown that speed-ups obtained reach around 90 for parallelization of an irregular adaptive integration code compared to maximum speed-ups of 98 for code without thread management at various levels of the divide-and-conquer tree. Results for various thread affinities are shown. The framework, for which the source code is presented, can be easily reused for any other divide-and-conquer application.

Keywords Divide-and-conquer · Parallelization · Adaptive integration · Intel Xeon Phi

1 Introduction

Recently the performance growth in HPC (high performance computing) systems has been achieved by packing more cores into traditional CPUs as well as incorporation of accelerators into computer nodes. The current top cluster MilkyWay-2 on the TOP500[1] list incorporates Intel Xeon E5-2692 12C CPUs running at

[1]http://top500.org.

P. Czarnul (✉)
Faculty of Electronics, Telecommunications and Informatics,
Gdansk University of Technology, Gdańsk, Poland
e-mail: pczarnul@eti.pg.gda.pl

© Springer International Publishing Switzerland 2016

J. Świątek et al. (eds.), *Information Systems Architecture and Technology: Proceedings of 36th International Conference on Information Systems Architecture and Technology – ISAT 2015 – Part III*,
Advances in Intelligent Systems and Computing 431, DOI 10.1007/978-3-319-28564-1_9

2.200 GHz with Intel Xeon Phi 31S1P coprocessors for the total of 3,120,000 cores and 33.8 PFlop/s performance while the second cluster Titan combines Opteron 6274 16C CPUs running at 2.200 GHz with NVIDIA K20x GPUs for the total of 560,640 cores and 17.59 PFlop/s performance. On the other hand, programming parallel systems has become more difficult and cumbersome due to the variety of APIs used at various levels. These include OpenMP (Open Multi-Processing) and OpenCL (Open Computing Language) for multicore CPUs, NVIDIA CUDA API, OpenCL and OpenACC for GPUs, MPI (Message Passing Interface) for integration of computations among cluster nodes. MPI can be used with OpenMP or NVIDIA CUDA for efficient computing from multiple processes using Multi-process service. Packages are offered for management of computations on several nodes using APIs such as OpenCL [1]. Intel Xeon Phi allows to run already existing codes written using standard APIs for paralel programming such as OpenMP and MPI [2]. Consequently, it is a perfect candidate for parallelization of computations among a large number of threads. Driven by this, the author proposed a framework based on OpenMP for easy parallelization of divide and conquer problems and investigated performance on Intel Xeon Phi.

2 Related Work

Several solutions have been proposed in the literature for parallelization of divide-and-conquer computations. Rugina and Rinard [3] proposed design and implementation of a compiler for automatic parallelization of divide and conquer codes for which subproblems access disjoint regions of dynamically allocated arrays. Paper [4] presents a solution for partitioning sequential C divide-and-conquer code and executes it in parallel. In [5], the author proposed a C++ framework for automatic parallelization of divide-and-conquer applications. The system cuts off branches of a potentially irregular problem tree and assigns these for computations to processes on idle nodes of a cluster. Work [6] takes up the problem of load balancing in divide-and-conquer problems being parallelized. The authors use dynamic group splitting and assign a certain number of processors to subtasks. The authors consider two strategies for coping with imbalance: repivoting and serialization. Implementation and parallel execution of divide-and-conquer algorithms is easily possible in Cilk Plus [7] which allows to extend the sequential code with constructs such as cilk_spawn and cilk_sync.

Work [8] lists various programming models and best practices for programming on the Xeon Phi platform. Paper [9] compares performance of various processors for sparse matrix vector (SpMV) and sparse matrix multiplication (SpMM).

Work [10] studies the performance of Xeon Phi for NAS benchmarks. Performance results are presented for varying numbers of threads for various settings of CPU affinity. The authors compared performance results of Xeon Phi to Xeon Sandy Bridge CPUs using either 16 or 32 threads for best results. For BT, CG, FT, IS, LU, and SP benchmarks Sandy Bridge exhibits better performance. For MG, Xeon Phi exhibited better performance than Sandy Bridge. It is concluded that many parallel loops are too short to be parallelized effectively among the many Phi threads. In [11], the authors compared performance of three different applications such as Fibonacci, NQueens and Cholesky factorization on Intel Sandy Bridge and Xeon Phi using three programming platforms: XKaapi, CilkPlus and OpenMP. The authors show that for computing a fine grained application—the n-th Fibonacci numer recursively, the best performance is obtained by XKaapi followed closely by CilkPlus and much slower OpenMP implementation. For NQueens the best performance is offered by XKaapi followed by OpenMP and CilkPlus. For Cholesky factorization, relative results are similar with almost identical performance of OpenMP and CilkPlus. In work [12] similar performance of CilkPlus and OpenMP is demonstrated for Dense Linear Algebra Factorization on Knights Ferry with slightly worse results for CilkPlus when running with two threads per core compared to slight improvement for OpenMP in this case. Paper [13] presents acceleration of short-range molecular dynamics simulations in a hybrid CPU-Xeon Phi environment. One 8-core Intel Xeon E5-2670 CPU clocked at 2.60 GHz and one 57-core Intel Knight Corner coprocessor were used. The parallel efficiency of 72.4 % was obtained for 57 threads on the co-processor. The overhead for various OpenMP constructs such as parallel for, barrier and reduction was measured in [14] along with memory bandwidths for different numbers of threads and affinities. The authors show that an OpenMP implementation of a Conjugate Gradient solver reaches the speed-up of over 70 compared to an Intel MKL based implementation with the speed-up of 80. In [15] the authors assessed the performance of an Intel Xeon Phi based platform compared to a Sandy Bridge system. Speed-up results for NAS paralel benchmarks range from 44.34 for FT up through 74.51 for SP up to 113.82 for EP on an Intel Xeon Phi.

3 Motivations and Proposed Solution

In view of the aforementioned developments, this paper proposes a flexible framework for parallelization of potentially irregular divide-and-conquer computations that manages threads and execution on the underlying hardware. Furthermore, it demonstrates the flexibility and usefulness of the framework and good performance of the OpenMP implementation utilizing multiple cores of the

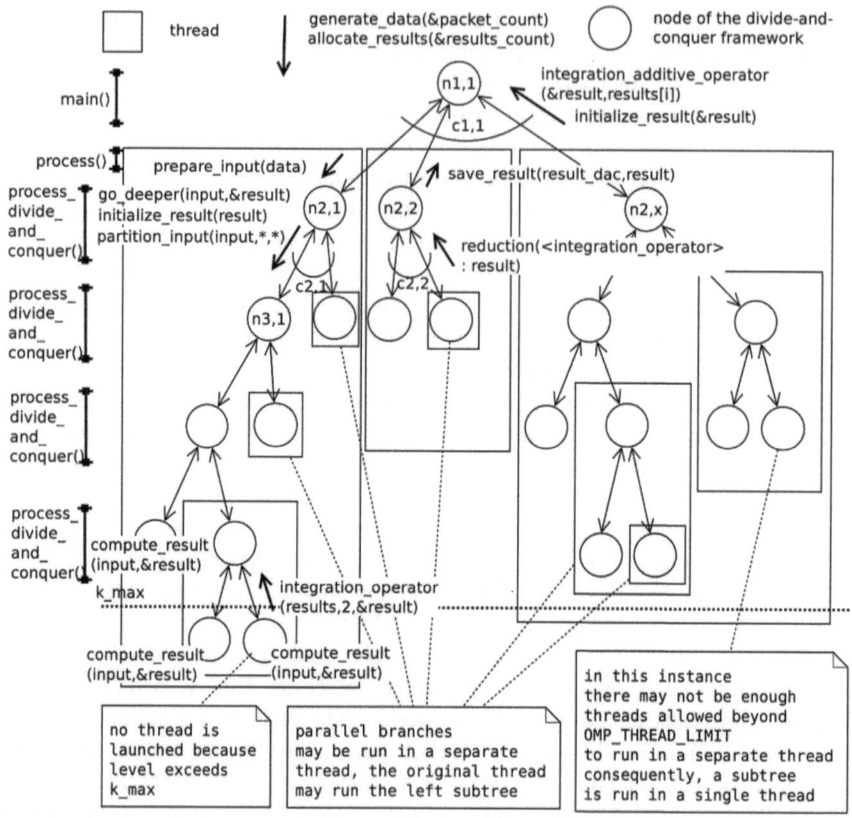

Fig. 1 Divide-and-conquer framework in OpenMP

Intel Xeon Phi in an easy way to achieve high scalability. In the divide-and-conquer approach, the initial problem is divided into a certain number of subproblems for which the same scheme is applied recursively. This generates a tree that generally may be imbalanced. Both the numbers of subproblems and depths of the tree in various parts may differ in which case it describes an irregular application [5]. The scheme and a model of the problem is shown in Fig. 1 with the following notation used: ni, j denotes the j-th node at level i where i starts with 1 at the root and j starts with 1 from the left side of the processing tree, ci, j denotes the number of subproblems generated by node ni, j.

Codes 1 and 2 present a parallel pseudocode using OpenMP for parallelization of branches by spawning threads that can be executed in parallel. While running multiple threads can exploit hardware such as cores in the Intel Xeon Phi, spawning too many threads may result in too high switching overheads and drop in speed-up. Because of that, the total number of threads is limited by setting an environment variable OMP_THREAD_LIMIT. On Intel Xeon Phi it is recommended to run at least 3 to 4 threads per physical core in order to achieve high performance although some codes may approach this limit with 2 threads per core [16]. The proposed C code framework features the following important methods (Fig. 1). Method main() (shown in detail in Code 1) initializes processing by generation of input data packets in function generate_data(&packet_count), allocates space for results in function allocate_results(&results_count) and launches threads at the highest level that subsequently fetch data packets in a critical section and process in a divide and conquer fashion. Then results are integrated. Data initialization and management code including function process(t_data data,t_result result) that prepares data for recursive execution in function prepare_input(data) and parses results afterwards in function save_result(result_dac,result). Method process_divide_and_conquer() shown in Code 2 first checks whether the current depth in the tree is not further from the root than threshold k_{max}—if so partitioning among threads might occur, otherwise computations are performed within the same thread. It executes function go_deeper(input,&result) that checks whether the problem given in input should be further divided into subproblems. This checking might produce a result at this level that might be used when the current subproblem is not further divided into subproblems. In case the problem is divided then a #pragma omp parallel block allows execution of subproblems in several threads (for clarity Code 2 shows this for two threads): each thread invokes the function recursively on its individual input with an increased depth $k + 1$, results from the threads are integrated, in this case using an OpenMP construct reduction (<integrationoperator>:result). It may also be implemented using private variables put into the lastprivate construct. In case threads may not be started in parallel due to a constraint set using OMP_THREAD_LIMIT then the function is invoked on the input data with an increased depth. What this causes is another possibility for running in multiple threads if newlevel $< k_{max}$ or partitioning of data and recursive execution in one thread otherwise.

Function main.c for the proposed solution (Code 1):

```
#define HIGH_LEVEL_THREAD_COUNT 60 #define PACKET_COUNT 10000
typedef struct { void *data; } t_data;
typedef struct { void *result;} t_result;
t_data *data; t_result *results;
main (int argc, char **argv) {
  // now the main processing loop using OpenMP
  int counter; int my_data; double result; int i; int packet_count; int results_count; int threadid;
  data=generate_data(&packet_count); // generate input data - specific to the problem
  results=allocate_results(&results_count); // allocate space for results
 omp_set_nested(1); omp_set_dynamic(0);
#pragma omp parallel private(my_data) firstprivate(threadid)
shared(counter) num_threads(HIGH_LEVEL_THREAD_COUNT)
  {
     threadid=omp_get_thread_num();
     do {
       // each thread will try to get its data from the available list
       #pragma omp critical
        { my_data=counter;counter++; } // also write result to the counter
        // process and store result -- this can be done without synchronization
        if (my_data<PACKET_COUNT) process(data[my_data],results[my_data]);
        // note that processing may take various times for various data packets
     } while (my_data<PACKET_COUNT);
  }
   // now just add results with one thread
   initialize_result(&result);
   for(i=0;i<PACKET_COUNT;i++) integration_additive_operator(&result,results[i]));
}
```

The implementation uses two parameters that control the degree of parallelism:

1. k_{max}—a constraint that specifies the maximum level (counting from the root) until which spawning thread teams is allowed. The rationale is that spawning threads for too small problems may not yield enough gain from parallelization and bring too much overhead.

2. max_thread_count—the total number of threads allowed to be run at any given time which can be specified by an environment variable OMP_THREAD_LIMIT.

Setting a small value of k_{max} may result in an insufficient number of threads to utilize the cores available in the accelerator. On the other hand, setting a higher value of k_{max} without other limitations may result in too many threads and considerable switching overheads. The problem considered in this paper for a given divide-and-conquer algorithm is to find such k_{max} and max_thread_count that would minimize the running time of the algorithm through breaking computations in the tree into reasonably sized subtrees that will be executed on separate cores. The global limit of max_thread_count is used to control the granularity of parallelization.

Recursive process_divide_and_conquer.c function parallelized using threads (Code 2):

```
dtype_result process_divide_and_conquer(dtype_input in-
put, int k) {
   dtype_result  result, results[2]; dtype_input in-
put1,input2; // inputs to threads
   int threadid,threadcount;
   if (k<k_max) { // checking for k_max -> generate new
threads
      if (go_deeper(input,&result)) { // go deeper
         initialize_result(result);
        // initialize input for threads
        input1=partition_input(input,1,2); in-
put2=partition_input(input,2,2);
        #pragma omp parallel private(threadid,threadcount)
firstprivate(input1,
           in-
put2,k)reduction(<integration_operator>:result)
num_threads(2)
        { threadid=omp_get_thread_num();
threadcount=omp_get_num_threads();
          if (threadcount==2) {
             if (threadid==0) { re-
sult=process_divide_and_conquer(input1,k+1);
            } else { re-
sult=process_divide_and_conquer(input2,k+1); }
          } else { //= one thread active here
            re-
```

```
sult=process_divide_and_conquer(input,k+1);}
          }
          return result;
        } else compute_result(input,&result);
      } else { // no new threads allowed because of the
k_max constraint
        if (go_deeper(input,&result)) { // go deeper
          // initialize input
          input1=partition_input(input,1,2); in-
put2=partition_input(input,2,2);
          result[0]=process_divide_and_conquer(input1,k+1);
          result[1]=process_divide_and_conquer(input2,k+1);
          integration_operator(results,2,&result);
        } else compute_result(input,&result);
      } return result;
    }
```

4 Experiments

4.1 Testbed Application and Environment

For the first and following tests, the framework described in Sect. 3 was filled in with code for numerical integration of function sin2(x)/x over a given range. Method go_deeper(input,&result) is implemented as follows: the initial range [x1, x2] is divided into two subranges [x1, (x1 + x2)/2] and [(x1 + x2)/2, x2] for which areas are computed as areas of trapezoids—a, a1 and a2 respectively. Method go_deeper(input,&result) returns fabs(a–a1–a2) > 0.000000001 i.e. if the difference between areas a and sum of a1 and a2 is large enough then [x1, x2] is further divided. Method partition_input(input, n, m) divides the initial range into two subranges for which computations are performed in parallel. integration_operator (results,n,&result) is implemented as reduction(+:r) in OpenMP. Method compute_result(input,&result) can use the previously computed value a.

One of key features of the Xeon Phi platform is vectorization. In our case, the compiler vectorized crucial loops in the code computing the integrate as a sum of rectangles over the computed range that gave the speed-up of 15.09 compared to non-vectorized code. FMA (Fused Multiply-Add) instructions are supported which operate on three registers out of which one acts as destination. The tests for which results are shown in Figs. 2, 3 and 4 were performed using a server with two Intel(R) Xeon(R) E5-2690 at 2.90 GHz CPUs with an Intel Xeon Phi 3120 coprocessor with 57 physical cores clocked 1.1 GHz with 6 GB of memory.

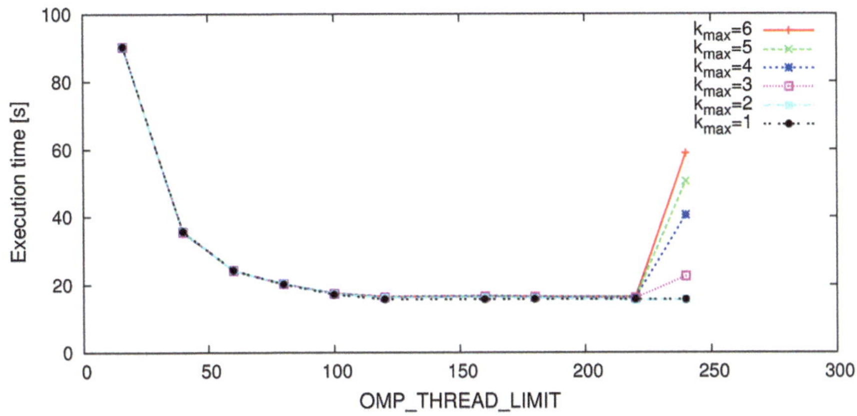

Fig. 2 Execution times for various values of k_{max} and max_thread_count set using OMP_THREAD_LIMIT, range [1, 100,000,000]

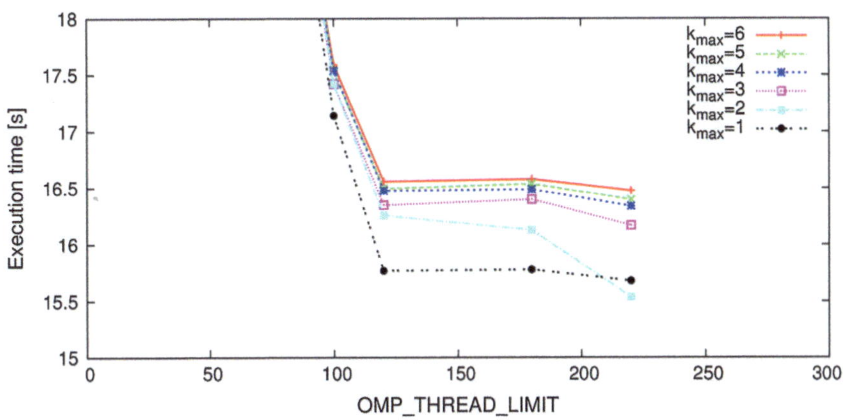

Fig. 3 Detailed view on execution times for various values of k_{max} and max_thread_count set using OMP_THREAD_LIMIT, range [1, 100000000]

4.2 *Execution Times and Overhead for Various* $\mathbf{k}_{max}$ *Values*

Figures 2 and 3 present execution times for the divide and conquer scheme using the testbed application and varying numbers of k_{max}. The latter denotes the maximum level number at which creation of OpenMP thread teams is allowed. Execution times are drawn against various values of max_thread_count—the maximum number of threads that are allowed within the application at any given time. This is imposed by setting environment variable OMP_THREAD_LIMIT.

The default value of thread affinity was used. From the figure, several conclusions can be drawn:

1. For a given k_{max}, the shortest execution times are obtained for the largest max_thread_count not exceeding 4 times the number of physical cores. Especially for larger k_{max}, allowing too many threads results in visible overhead.
2. For various values of k_{max} and identical values of max_thread_count not exceeding 4 times the number of physical cores the execution times are very similar. A more detailed plot shown in Fig. 3 reveals small differences with larger execution times for larger values of k_{max}. This is expected because allowing to spawn thread teams at more levels might result in a larger total overhead for thread creation. While the overhead is indeed larger for larger k_{max}, differences are small. This means that allowing thread teams for larger k_{max} can be set by default at a small cost but with the potential advantage of better load balancing if differences between execution times of threads spawned at lower levels are significant.

4.3 Various Data and Thread Management Methods

While the proposed framework gives the flexibility of the divide-and-conquer scheme with automatic management of parallelization and the number of threads constrained by the value of OMP_THREAD_LIMIT, it imposes overheads. It is necessary to assess these overheads. Firstly, a version with adaptive integration with threads launched at the beginning of computations was developed. In this case, threads are launched in function main(). This means that this version does not handle load balancing within each branch of the divide-and-conquer tree but balances whole branches started from the root. As such, it is used here as a reference point for performance. Figure 4 shows a comparison of speed-ups of this version against the flexible divide-and-conquer with threads managed at runtime described in Sect. 3 and suggests that the adaptive code exhibits the speed-up from 98.3 with threads created for top-level branches. There is a slight drop to 91.5 and 87.05 for the divide-and-conquer version with $k_{max} = 1$ and $k_{max} = 6$ respectively. This gives an acceptable decrease from 7 to 11.5 % respectively. In return, the programmer gets a flexible programming scheme and does not need to deal with assignment of computations to particular threads.

Furthermore, it should be noticed that the adaptive quadrature framework introduces imbalance of execution times among threads by its very nature. Execution times of threads may differ significantly for various subranges because various accuracy might be needed in the latter. As a reference, the author has implemented a static code which uses integration of a subrange by adding areas of 10^7 rectangles into which each subrange is divided. The initial subrange was divided into 10^4 subranges in function main(). Each thread would fetch successive subranges from the initial set, compute result and fetch another subrange for computations. For integration of the same range [1, 10^8] the flexible divide-and-conquer framework offers respectable speed-ups of up to 91.5 for $k_{max} = 1$ and 87 for $k_{max} = 6$ while the best speed-up for the static non-adaptive version is around 108.

Table 1 shows execution times for a selected test case using KMP_AFFINITY values such as compact, scatter, balanced that determine how threads are scheduled on Xeon Phi cores [17]. This particular test was run on an Intel Xeon Phi 5100 series model with 60 physical cores. The best times are obtained for balanced followed by scatter and compact with relatively small differences. Using another test case the Intel Xeon Phi 3120 finished computations in 10.48 s (228 threads) compared to a dual CPU E5-2690 2.90 GHz which finished in 15.83 s (32 threads).

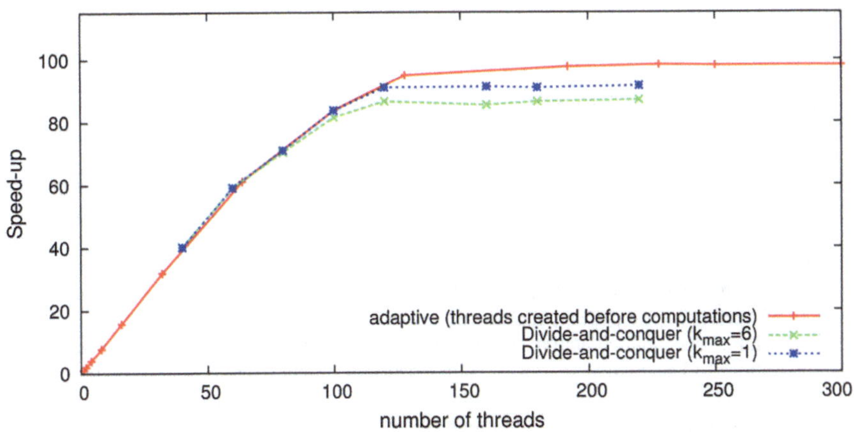

Fig. 4 Comparison of speed-ups for adaptive integration and either statically assigned or dynamically managed threads for computations

Table 1 Comparison of execution times [s] for various thread affinities, 228 threads

k_{max}	Compact	Scatter	Balanced
6	34.4035	33.3911	33.2698
5	34.3015	33.3911	33.2698
4	34.2152	33.2981	33.2277
3	34.0646	33.2075	33.0865
2	33.4953	33.0805	32.8387
1	31.8776	31.7029	31.6990

5 Conclusions and Future Work

The paper presented an OpenMP based framework for parallelization of divide-and-conquer applications benchmarked on an Intel Xeon Phi. The latter allowed obtaining short execution times and high speed-ups thanks to many cores and vectorization. It was confirmed in real experiments that using the framework for parallelization of an adaptive and irregular numerical integration application resulted in speed-ups around 90 on a single Intel Xeon Phi. Future work includes testing more parallel applications.

Acknowledgments The work was supported by Intel within grant "Intel Phi Parallel Processing Lab". The author wishes to thank Intel, especially Marek Zmuda, Marek Piosik and Robert Bard from Intel for provision of Xeon Phi equipment, literature and needed support.

References

1. Czarnul, P., Rosciszewski, P.: Optimization of execution time under power consumption constraints in a heterogeneous parallel system with gpus and cpus. In: Distributed Computing and Networking, pp. 66–80. Springer Berlin (2014), Volume 8314 of LNCS
2. Jeffers, J., Reinders, J.: Intel Xeon Phi Coprocessor High Performance Programming. Newnes, New South Wales (2013)
3. Rugina, R., Rinard, M.: Automatic parallelization of divide and conquer algorithms. In: Proceedings of the Seventh ACM SIGPLAN Symposium on Principles and Practice of Parallel Programming, pp. 72–83. PPoPP '99, New York, NY, USA, ACM (1999)
4. Freisleben, B., Kielmann, T.: Automatic parallelization of divide-and-conquer algorithms. In: Parallel Processing: CONPAR 92—VAPP V, pp. 849–850. Springer, Berlin (1992), Volume 634 of Lecture Notes in Computer Science
5. Czarnul, P.: Programming, tuning and automatic parallelization of irregular divide-and-conquer applications in Dampvm/DAC. Int. J. High Perform. Comput. Appl. **17**, 77–93 (2003)
6. Eriksson, M.V., Keßler, C.W., Chalabine, M.: Load balancing of irregular parallel divideand-conquer algorithms in group-spmd programming environments. In: ARCS Workshops, pp. 313–322. GI (2006), Volume 81 of LNI
7. Intel: Intel cilk plus language specification (2010) ver. 0.9. http://www.cilkplus.org/sites/default/files/open_specifications/cilk_plus_language_specification_0_9.pdf
8. Michaela, M., Byckling, M., Ilieva, N., Saarinen, S., Schliephake, M., Weinberg, V.: Best practice guide intel xeon phi v1.1, PRACE, 7 Capacities (2014)
9. Saule, E., Kaya, K., Çatalyürek, Ü.V.: Performance evaluation of sparse matrix multiplication kernels on intel xeon phi. CoRR abs/1302.1078 (2013)
10. Ramachandran, A., Vienne, J., der Wijngaart, R.A., Koesterke, L., Sharapov, I.: Performance evaluation of nas parallel benchmarks on intel xeon phi. In: IEEE ICPP, pp. 736–743 (2013)
11. Lima, J.V., Broquedis, F., Gautier, T., Raffin, B.: Preliminary experiments with xkaapi on Intel xeon phi coprocessor. In: Symposium on Computer Architecture and High Performance Computing
12. Eisenlor, J., Hudak, D., Tomko, K., Prince, T.: Dense linear algebra factorization in OpenMP and Cilk Plus on Intel MIC: development experiences and performance analysis. In: TACCIntel Highly Parallel Computing Symposium (2012)

13. Wu, Q., Yang, C., Tang, T., Xiao, L.: Mic acceleration of short-range molecular dynamics simulations. In: Proceedings of the First International Workshop on Code OptimiSation for MultI and Many Cores. COSMIC ’13, New York, NY, USA, ACM (2013) 2:1–2:8
14. Cramer, T., Schmidl, D., Klemm, M., Mey, D.: Openmp programming on intel xeon phi coprocessors: An early performance comparison. In: Proceedings of the Many-core Applications Research Community Symposium at RWTH Aachen University, pp. 38–44 (2012)
15. Schmidl, D., Cramer, T., Wienke, S., Terboven, C., Müller, M.: Assessing the performance of openmp programs on the intel xeon phi. In: Euro-Par 2013 Parallel Processing, pp. 547–559. Springer, Berlin (2013), Volume 8097 of LNCS
16. Reinders, J.: An overview of programming for intel xeon processors and intel xeon phi coprocessors, Intel. https://software.intel.com/sites/default/files/article/330164/an-overview-of-programming-for-intel-xeon-processors-and-intel-xeon-phi-coprocessors_1.pdf (2012)
17. Green, R.W.: Openmp* thread affinity control. compiler methodology for intelR mic architecture. https://software.intel.com/en-us/articles/openmp-thread-affinity-control (2012)

Part III
Knowledge Discovery and Data Mining

Information Technology as an Indirect Factor Supporting Knowledge Management in Organization—Mediating Role of Organizational Structure

Katarzyna Tworek, Katarzyna Walecka-Jankowska and Janusz Martan

Abstract The paper is devoted to analyze the role of IT in managing knowledge in organization. Organizational structure is presented as a factor through which IT influences knowledge management and facilitates its better usability in organization. Empirical studies concerning this topic are presented. First part of studies describes how the organic-type organizational structure influences knowledge management in organization. Then, second part is devoted to analyze IT as a factor facilitating reorganization into more organic-type organizational structure. Conclusion of the paper is that IT influences organization ability to reorganize into organic-type structure and because of that IT can be perceived as a factor affecting knowledge management in organization also in indirect way.

Keywords Information technology · Knowledge management · Organizational structure

1 Introduction

Rapid access to knowledge is a critical factor of success for many organizations [1]. Therefore knowledge management (KM) is an important field of studies and factors that can influence it are worth analyzing. Information technologies (IT) is perceived as one of this factors. Along with the changes in the world at the turn of the century, it almost became a necessity to use IT in KM for the organization benefits. Therefore, it is important to determine exactly how IT can influence KM in organization. Thus, it will be the main subject of this paper.

K. Tworek (✉) · K. Walecka-Jankowska · J. Martan
Wrocław University of Technology, Wrocław, Poland
e-mail: katarzyna.tworek@pwr.edu.pl

K. Walecka-Jankowska
e-mail: katarzyna.walecka-jankowska@pwr.edu.pl

J. Martan
e-mail: janusz.martan@pwr.edu.pl

© Springer International Publishing Switzerland 2016

J. Świątek et al. (eds.), *Information Systems Architecture and Technology: Proceedings of 36th International Conference on Information Systems Architecture and Technology – ISAT 2015 – Part III*, Advances in Intelligent Systems and Computing 431, DOI 10.1007/978-3-319-28564-1_10

In the literature it is underlined that IT is connected with KM. It is a tool to acquire, transfer and codify knowledge. Authors of these paper would like to underline another quality of IT—it facilitates reorganization into more organic-type organizational structure indirectly (but hugely) impacting knowledge management processes. Therefore—IT can be seen not only as a tool supporting KM but also as a facilitator of changes that support KM in organization.

2 Knowledge Management (KM) and Information Technology (IT)

2.1 Knowledge Management (KM)

Knowledge management (KM) has no clear definition, which would be widely accepted by both practitioners and theorists. Definitions of KM presented in the literature often focus on the areas of KM in organizations—it is process approach—based on consulting company experience, the main emphasis is on KM process: acquisition, transfer, codification (for this approach characteristic is logic and ordering, but knowledge is identified as an information). The emphasis in process approach is not always located on all processes—some are focused on the processes of transfer and sharing, some on the possibility of acquiring knowledge. There are two more—resource approach (shows how organizations generate and use knowledge, it is based on strategic approach and knowledge is identified as most important resource, which should be managed to be competitive advantage) and 'Japanese' (concentrates on creating and breading knowledge in social interaction process between tacit and explicit knowledge; so the most important thing in this approach is definition of tacit and explicit knowledge). The synthesis of tree approaches to KM was use in research (according to resource approach model shows how organizations generate and use knowledge; simultaneously knowledge is treated as specific and distinctive resource so demands different treatment; and processes, but does not equate knowledge with information; from Japanese approach were taken two kinds of knowledge: tacit and explicit. And model contains actions directed to sharing of tacit knowledge and codification of explicit knowledge).

KM processes follows one another with input and output. The processes are: acquisition (creation and location outside the organization), codification, transfer, utilization and storage knowledge in organization. They form a logical sequence of events: the acquisition or development of knowledge followed by recording and codification of good practice (Fig. 1).

New knowledge is the first step in the cycle of KM, which is oriented towards creation of mechanisms driving the cycle of creating knowledge, which is valuable to the organization, employees and customers. Second process—codification provides access to knowledge, thereby increasing the degree of dissemination of

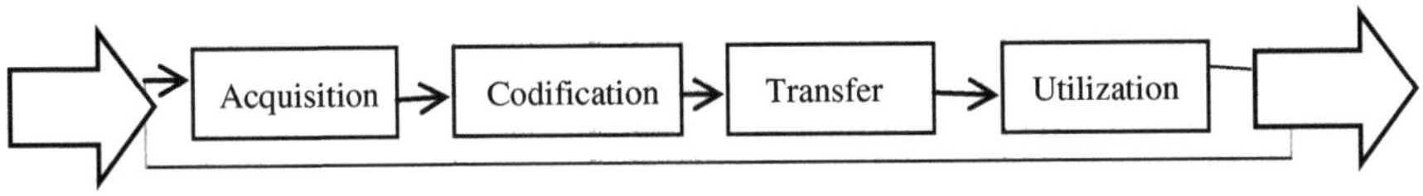

Fig. 1 Processes of knowledge management

knowledge by transferring it. The last element is the use of knowledge that occurs as a result of increasing competence of employees and the organization as a whole. This leads to the closing of the cycle, as an organization with the new knowledge can see new possibilities of its creation and development.

2.2 *Information Technology (IT)*

Information technology (IT) is defined as application of computers and telecommunications equipment to store, retrieve, transmit and manipulate data [2]. In today's world, with emerging economy, the increase of importance of information technology use in organizations is caused primarily by the need of knowledge and information management. Along with the changes in the world at the turn of the century, it almost became a necessity to have the knowledge and skills that allow the use of information technology for the company benefits.

The most approved method of characterizing IT in organization is the analysis of its dissemination but it is important to analyze IT with a use of variables that allow to specify how many employees actually use IT in their work [3]. That was the genesis of the concepts of IT acceptance. It has allowed to take into consideration the imbalance between the number of employees who have access to IT and the number of those actually using it. IT acceptance, understood as a demonstrable willingness within a user group to employ IT for the tasks it is designed to support [4], will be a variable used in this publication to characterize IT in organizations.

2.3 *Organizational Structure—Factor Connecting IT and KM*

The necessity of effective KM, recognized as a strategic corporate asset, is forcing changes in a way of organization work. The transformation of modern organizations in knowledge-based ones is its consequence. This requires actions undertaken in different areas of the organization, aimed (on the highest level of generality) at providing a competitive advantage through the development of knowledge and skills. It happens because of organizational culture that supports the acquisition and sharing of knowledge, the design of structures and procedures to ensure the collection and use of knowledge, the acquisition of knowledge from the environment market and use of IT systems, which allow to codify, collect and share knowledge between members of the organization [5].

Organizational structure has the key role in understanding relation between IT and KM. The organizational structure is defined by 5 dimensions formalization, centralization, specialization, hierarchy, standardization. According to Lee and Grover [6], Liao [7], Chen and Huang [8] and Gonzales eta al. [9] three dimensions: formalization, centralization and integration are related to KM process. Literature studies indicate that structural features, which approximate it to more organic (flatter, decentralized) are more effective for KM [8–10]. Since KM depends on social interactions and on the flow of knowledge between individuals and departments, the organizational structure exerts great influence [8, 9, 11]. Activities within the subsequent KM processes have a direct relations with the dimensions of the organizational structure, especially the formalization, standardization and centralization level. It is important to point out that in order to acquire and develop knowledge, there are mergers of departments and individuals taking place in the organization. The mergers may be temporary and reorganize structure towards more flexible forms. Employees are expected to be intellectually committed. Subjective approach to employees has to be encouraging, should mobilize them to ingenuity and promote self-organization at the same time. The knowledge acquired and developed at an early stage must then be structured and stored—it is the purpose of codification, which allows to understand the relationships between data, its identification, documentation of rules governing data management. It also ensures the integrity and relevance of the data—basing on the principles and procedures. Codification of knowledge and mapping its sources can affect the reduction in the level of formalization—eliminating data redundancy, allowing access to data in electronic form by many people at the same time. On the other hand when organizational processes are standardized and documented, knowledge replication is easier (such as hand-offs and specs).

Transfer of knowledge is aimed at reducing the incidence of not using knowledge by the organization. It requires the identification of areas from which it can be freely acquired and clear rules of its location. The organizations which adopt flexible, increasingly flat organizational forms with fewer hierarchical levels which not only allow but also and teamwork. This makes it possible for employees to take better advantage of their competencies, generating organizational routines and increasing the value of their contributions thanks to the freedom of action they are given [10]. The ability of sharing and distributing knowledge resources across functional boundaries enables the firm to fundamentally change its business processes. The sharing of knowledge resources not only facilitates cross-functional interaction but also allows the sharing of knowledge repositories among process participants, thereby allowing greater collaboration and understanding of the entire process rather than having fragmented parts of the process [1]. Transfer of knowledge includes two forms: formal—e.g. IT can be used to transfer explicit knowledge (workflow) and informal, which is based on the assumption that the simplest and most effective form of knowledge transfer is direct and free contact between employees. In organizations which manage their knowledge it is common to observe shorter communication channels to facilitate the rapid transfer of knowledge. Knowledge sharing and the availability of required knowledge empower employees at the lower level of organizational hierarchy may encourage them to participate in more decision making activities than they would otherwise—this tendency would result in a more decentralized organizational structure [1].

Some authors concentrate on direct relation between KM and IT—i.e. Gonzales et al. [9] are of the opinion that distribution of knowledge with respect to the portion of the explicit knowledge are supported and expanded by the IT tools. Usually role of IT is under consideration especially with knowledge transfer. Organizations depend on IT to store, formalize and distribute the explicit knowledge [12]. But transfer of knowledge includes two forms: formal—where indeed IT can be used to transfer explicit knowledge (workflow), but also informal—which is based on the assumption that the simplest and most effective form of knowledge transfer is direct and free contact between employees [2, 5]. To better facilitate this kind of knowledge transfer in organization—the organic-structure is needed. It enables the flow of knowledge between individuals, organizational departments and other organizations, also leading to creation of mechanisms for the integration of individuals [11]. Thus, this paper considers IT as a facilitator of reorganization into more organic-type structure for KM process.

Moreover, organic-structure creates an environment conducive to the acquisition and creation of knowledge because of low level of formalization and centralization. Employees have the opportunity for direct contact with one-another, often using a variety of tools (regardless of their position or organizational level)—which once again suggest a big role of IT in facilitating knowledge management in organization.

3 Information Technology (IT) Influence on Knowledge Management (KM)—Empirical Studies

Empirical research has been done in two parts. First part concerns relation between organizational structure and KM. Second part describes relation between IT and organizational structure.

3.1 Organizational Organic Structure as a Factor Influencing KM

The general aim of the research was to define the determinants of the organizational innovation, with a particular emphasis on the KM processes. During studies 105 organisations operating in Poland were examined (they were different in terms of size, industry and ownership structure). One questionnaire was sent to each surveyed organization with the request that a person with a broad view of the whole organization (i.e. CEO, management team, quality specialist etc.) fill it in. The results should be regarded as a kind of pilot studies because the sample of the surveyed organisations was not representative.

To investigate the results of the relation between KM processes, and organizational structure, the key variables were defined. KM was described by degree of 4 knowledge processes: acquisition and creation (16 positions in scale), codification (18 positions), transfer (11 positions) and utilization (2 positions) of knowledge. Additionally, Cronbach's α was calculated for this scales—it came back as 0.886 (acquisition and creation), 0.934 (codification), 0.843 (transfer) and 0.889 (utilization) which indicates very high internal reliability of all four scales. Set of statements and detailed information were published in [13]). The key variables for organizational structure dimensions were defined: degree of specialization, formalization, standardization, centralization and hierarchy coded in 10 items (Table 1).

Firstly, correlation analysis using Pearson Coefficient was done. Analysis has indicated that there is indeed the correlation between knowledge processes and degree of formalization, standardization, hierarchy, centralization and specialization, which may indicate that with the changes of organizational structure into more organic-type one, knowledge processes are more applied. Stepwise regression analysis was performed, results are presented in Table 2 (including statistical significance of each variable from the model).

Models were obtained for acquisition ($F(5,99) = 5.861$, $p < 0.01$), codification ($F(5,99) = 5.474$, $p < 0.01$) transfer ($F(5,99) = 4.436$, $p < 0.001$) and utilization ($F(5,99) = 5.384$, $p < 0.001$). Regression analysis has shown that increased level of characteristics connected with organic-type structure causes a significant change in variables characterizing KM processes. It confirms that dimensions of organizational structure are an important features for KM processes.

Table 1 Organizational structure items

Hierarchy	Centralization	Formalization	Specialization	Standardization
Number of managerial positions is high	Decisions concerning strategic problems are made on highest hierarchy level	Number of organizational documents is high	Tasks are simple and repetitive	Number of standard procedures is high
Number of hierarchy levels is high	Operational decisions are made in place of the problem	Degree of documents details is high	Number of jobs is high	Adherence to the specific procedures is required

Table 2 Regression analysis between KM processes and characteristics of organizational structure

Model		Non-standardized coefficients		Standardized coefficients	T	Signif.
		B	Standard error	Beta		
Acquisition and creation	(Constant)	1.371	0.424		3.232	0.002
	Formalization	−0.042	0.084	−0.061	−0.506	0.614
	Standardization	0.467	0.105	0.547	4.444	0.000
	Hierarchy	−0.009	0.078	−0.013	−0.117	0.907
	Specialization	−0.240	0.088	−0.308	−2.715	0.008
	Centralization	0.145	0.088	0.147	1.646	0.103
Codification	(Constant)	1.248	0.490		2.548	0.012
	Formalization	−0.001	0.097	−0.001	−0.010	0.992
	Standardization	0.437	0.121	0.446	3.602	0.000
	Hierarchy	0.010	0.090	0.013	0.115	0.909
	Specialization	−0.308	0.102	−0.345	−3.020	0.003
	Centralization	0.254	0.102	0.225	2.501	0.014
Transfer	(Constant)	1.393	0.449		3.103	0.002
	Formalization	−0.123	0.089	−0.172	−1.385	0.169
	Standardization	0.472	0.111	0.538	4.250	0.000
	Hierarchy	−0.025	0.082	−0.036	−0.309	0.758
	Specialization	−0.117	0.093	−0.147	−1.257	0.212
	Centralization	0.144	0.093	0.143	1.551	0.124
Utilization	(Constant)	0.876	0.687		1.275	0.205
	Formalization	−0.019	0.136	−0.017	−0.141	0.888
	Standardization	0.600	0.170	0.438	3.525	0.001
	Hierarchy	0.026	0.126	0.023	0.208	0.836
	Specialization	−0.433	0.143	−0.346	−3.026	0.003
	Centralization	0.388	0.142	0.246	2.726	0.008

3.2 *IT as a Factor Influencing Organic Organizational Structure*

Research concerning relation between IT and organizational structure were conducted in 2012 (as a part of PhD Thesis [14]). 109 organizations were examined (they were different in terms of size, industry and ownership structure). All of these organization have implemented IT systems (they were different in terms of dissemination, user acceptance and type—construed as number and nature of implemented IT function) and claimed that IT is used, among others, to support organization learning.

To investigate the results of the relation between IT and organizational structure, the key variables describing 4 organizational structure dimensions were defined:

hierarchy, centralization, formalization (and standardization within this item), specialization. Subjective level of each variable was calculated as the arithmetic mean of grades given to each item describing one of 4 variables. Items were put in form of statements and they were rated by the respondents with Likert scale (5 items scale) (list of items—statements—for each variable is included in Table 3). It is worth noting that Cronbach's α was 0.641 and higher for every variable, which indicates a high internal reliability of the scales and measurements.

IT was described by degree of IT acceptance in organization. Subjective level of IT acceptance was calculated based on a set of statements prepared on basis of IT acceptance measuring method proposed by Davis—set of statements and detailed information were published by Tworek and Martan [3]. Additionally, Cronbach's α was calculated for this scale—it came back as 0.905 which indicates very high internal reliability of scale.

After confirming internal reliability of each scale and normal distribution for each variable, correlation and regression analysis was done. Results are presented in Table 4. Firstly, correlation analysis using Pearson Coefficient was done. Analysis has indicated that there is indeed the correlation between IT acceptance and hierarchy complexity, degree of centralization, formalization and specialization, which may indicate that with increase of IT acceptance, the organizational structure changes into more organic-type one (hierarchy is less complex and centralization, formalization and specialization degree is smaller).

Secondly, stepwise regression analysis was performed. Fitting models were obtained for hierarchy ($F(4,99) = 14.585, p < 0.001$), formalization ($F(3,100) = 8.357$,

Table 3 Items of organizational structure dimensions

Hierarchy	Centralization	Formalization (+Standarization)	Specialization
Organizational structure is complicated	Decisions concerning response to environment change are made on highest hierarchy level	Number of organizational documents is high	Tasks are simple and repetitive
Number of hierarchy levels is high	Decisions concerning intellectual capital are made on highest hierarchy level	Degree of documents details is high	Number of production workers is high
Number of direct subordinates of managers is high	Decisions concerning manufacturing technologies are made on highest hierarchy level	Number of regulations concerning ways of accomplishing tasks is high	Number of non-production works is high
Managers will not be able to manage more subordinates then they are managing now	Operational decisions are made in place of the problem	Regulations concerning employees communication are formal and restricted	Task rotation is common

Table 4 Regression analysis between organizational structure and IT acceptance

Independent variable in regression models		Dependent variable in regression model			
		1st model: hierarchy	2nd model: centralization	3rd model: formalization	4th model: specialization
Information technology acceptance	B	−1.133	−0.489	−0.481	−0.426
	Standardized error	0.193	0.150	0.201	0.129
	Beta	−0.538	−0.356	−0.248	0.277
	T-student test	−5.884	−3.262	−2.396	−3.303
	Significance	0.001	0.002	0.019	0.001
	Significant element of regression model	Yes	Yes	Yes	Yes

$p < 0.001$) and specialization (F (4,99) = 23.145, $p < 0.001$). IT acceptance proved to be a significant independent variable with every obtained model (Table 4). IT acceptance proved to be the only significant independent variable in Model obtained for centralization (F (1,102) = 9.690, $p < 0.05$). This model cannot be characterized as fitting—obtained predictors explained only 7.8 % of the variance of the dependent variable. Regression analysis has shown that with increase of IT acceptance there is a significant change in each variable characterizing organizational structure. Hierarchy complexity declines, as do the degree of centralization, formalization (standardization) and specialization. It confirms that organizational structure becomes more organic-type one.

Conducted research has shown that IT indeed impacts organizational structure and contributes to reorganizing to structure less complex, with lower degree of centralization, formalization and specialization. It can be concluded that the research provides evidence that in fact there is an increase of self-reliance among employees using IT, which contributes to the increase of the average span of management and results as lower centralization level and less complex structural hierarchy in organizations with highest IT acceptance. However, centralization, which is getting lower with increasing IT acceptance can be primarily the evidence of incensement of information synergy on the lower levels of management—the effect most desired in organizations trying to enhance their knowledge management processes. This increasing of IT acceptance leads to the situation, in which employees are in fact better equipped to make the right decisions in place, in which the problem arises and results with lower degree of centralization. Therefore, it can be concluded that increasing IT acceptance can in fact contribute to reorganizing structure into more organic-type one.

4 Conclusions

Presented research clearly confirms that IT influences organizational structure and facilitates reorganization into more organic-type structure. It is consistent with most common opinion among authors—that IT reorganizes the communication channels within the organization (making it easier for employers to communicate directly) and contributes to several changes in organizational structure. Looking at them from KM point of view and emphasizing those consistent with reorganization into more organic-type structure, those changes listed below can be seen as the most important ones [15, 16]:

- Reduction of the number of administrative employees and middle management employees mainly caused by a more efficient exchange of information.
- Increase of self-reliance due to better access to information among employees using IT, which contributes to the increase of the average span of management.
- Increase of information synergy on the lower levels of management, which leads to the situation, in which employees are better equipped to make the right decisions in place, in which the problem arises.
- Willingness among managers to delegate decisions to lower levels of the hierarchy due to the fact that employees acquire additional skills and knowledge while using IT.
- Emergence of intelligent formalization effect, which allows employers to communicate and exchange information in codified way, which at the end contributes to unification and organization of documents and procedures.
- Reduction of negative effects of specialization due to work enrichment and increased self-reliance of employees.

As it was mentioned at the beginning of the paper and confirmed by presented research—the organic-type structure and changes in organization that are consistent with it (listed above) are contributing to a better utility of KM in organization.

References

1. Liao, C., Chuang, S.H., To, P.L.: How KM mediates relationship between environment and organizational structure. J. Bus. Res. **64**, 728–736 (2011)
2. Crowston, K.: Processes as theory in information systems research. In: Arlborg, Niemcy, International Working Conference on the Social and Organizational Perspective on Research and Practice in Information Technology, pp. 149–164 (2000)
3. Tworek, K., Walecka-Jankowska, K., Martan, J.: Structure reorganization due to IT information functions support for knowledge management. China-USA Bus. Rev. **14**(4) (2015)
4. Dillon, A.: User Acceptance of Information Technology. In: W. Karwowski (Ed.) Encyclopedia of Human Factors and Ergonomics. Taylor and Francis, London (2001)
5. Walecka-Jankowska, K.: Wpływ zarządzania wiedzą na innowacyjność, PhD Thesis. Wroclaw University of Technology, Wroclaw (2011)

6. Liao, S.H.: Knowledge management technologies and applications—literature review from 1995 to 2002. Expert Syst. Appl. **25**(2), 155–164 (2003)
7. Ramezan, M.: Intellectual capital and organizational organic structure in knowledge society. Int. J. Inf. Manage. **31**, 88–95 (2011)
8. Chen, C.J., Huang, J.W.: How organizational climate and structure affect knowledge management: the social interaction perspective. Int. J. Inf. Manage. **27**(2), 104–118 (2007)
9. Domínguez Gonzalez, R.V., Martins, M.F.: Knowledge management: an analysis from the organizational development. J. Technol. Innov. **9**(1), 131–147 (2014)
10. Claver-Cortes, E., Zaragoza-Saez, P., Pertusa-Ortega, E.: Organizational structure features supporting knowledge manage KM processes. J. KM **11**(4), 45–57 (2007)
11. Chen, C.J., Huang, J.W., Hsiao, Y.C.: KM and innovativeness: The role of organizational climate and structure. Int. J. Manpow. **31**(8), 848–870 (2010)
12. Spanos, Y., Prastacos, G., Poulymenakou, A.: The impact of information and communication technologies on firm structure: evidence from an economy in transition. In: ICMIT, pp. 573–579 (2000)
13. Zheng, W., Yang, B., Mclean, G.B.: Linking organizational culture, structure, strategy, and organizational effectiveness: Mediating role of knowledge management. J. Bus. Res. **63**(7), 763–771 (2010)
14. Tworek, K., Martan, J.: Suitability analysis of a new method for measuring information technology acceptance. In: Kowal, J. (ed.) Communication and Information Technology in Society. Information and Communication Technologies (ICT) in Management, vol. 2, pp. 133–148. Newcastle upon Tyne: Cambridge Scholars Publishing, Cambridge (2015)
15. Jones, G.R., Dewett, T.: The role of information technology in the organization: a review, model, and assessment. J. Manag. **27**, 313–346 (2001)
16. Tworek, K.: Wpływ technologii informacyjnych na struktury organizacyjne przedsiębiorstw. PhD Thesis, Wroclaw University of Technology, Wroclaw (2014)

RT-PLG: Real Time Process Log Generator

Bernardo Nugroho Yahya, Yohanes Khosiawan, Woosik Choi and Ngoc Anh Dung Do

Abstract Streaming process mining has been rising as an emergent tool to analyze industrial practices. Obviously, the advance of streaming process mining requires the availability of a suite of real-world business processes and the execution logs in the real time manner. Literally, it is hard to obtain it. This paper aims to develop a real time process log generator for the usage of streaming process mining tool. The real time process log generator (RT-PLG) is constructed in an independent tool. Afterward, the RT-PLG is utilized to generate a synthetic log for streaming process mining. The tool has been evaluated using an existing simulation model.

Keywords Streaming process mining · Process log generator · Real time system

1 Introduction

Process mining has emerged as a way to analyze system and its actual usage based on the event logs they produce [1]. The goal of process mining is to extract useful knowledge such as process models, organizational models, etc. from the logs, which are commonly available in information systems [2, 3]. Most of process mining applications have focused on historical data and less concerned on real time

B.N. Yahya (✉) · W. Choi
Industrial and Management Engineering, Hankuk University of Foreign Studies, Oedae-ro 81, Mohyeon-myeon, Cheoin-gu, Yongin 449-791, South Korea
e-mail: bernardo@hufs.ac.kr

W. Choi
e-mail: dntlrdl@daum.net

Y. Khosiawan · N.A.D. Do
Mechanical and Manufacturing Engineering, Aalborg University, Fibigerstraede 16, 9000 Aalborg, Denmark
e-mail: yok@m-tech.aau.dk

N.A.D. Do
e-mail: ngoc@m-tech.aau.dk

© Springer International Publishing Switzerland 2016

J. Świątek et al. (eds.), *Information Systems Architecture and Technology: Proceedings of 36th International Conference on Information Systems Architecture and Technology – ISAT 2015 – Part III*, Advances in Intelligent Systems and Computing 431, DOI 10.1007/978-3-319-28564-1_11

data. In real time data, generated events include properties that are constantly changing and affected by time. These properties differ from traditional process mining containing persistent data. Hence, process mining in respect to streaming data, called as streaming process mining, has emerged.

There are many works on streaming process mining [4–7]. Most of the works attempt to solve the problem on discovering a good process model. Apparently, there is less work on process log generator that works for real time analysis. The work of Burattin and Sperduti has considered developing a process log generator based on Petri net without having a real time data [8, 9]. Some simulators [10, 11], or many called as simulation tool, usually provide the result of simulation without storing the detail event log. One tool called BP simulator models a process in the web and executes the model in a real time manner [12]. However, the simulator shows the event log only for data analysis and disregards the process perspective. Since the intention of process mining starts from real time log, it is necessary to build a real time process log generator to enhance and improve any streaming process mining algorithm.

This study aims to develop a real time process log generator (RT-PLG). This tool is obviously necessary to mimic the real world in real time manner. In addition, it raises some challenges on constructing process mining approach successively on forwarded data thereby updating the internal state. The development of this tool is beneficial for, in particular, the current issues on streaming process mining. As for verification, a streaming process discovery and process discovery using event logs are shown in this study.

The organization of this paper is as follows. Section 2 describes the streaming process mining framework. The implementation result is shown in Sect. 3. It explains the implementation of the proposed approach including real time process model discovery for future work. Finally, Sect. 4 concludes this paper.

2 Proposed Method

In this section, a real time process log generator framework is described. The framework shows how the real time data can be generated and how the real time process can be discovered. As shown in Fig. 1, it consists of five major steps: process model preparation, process model parsing, real time process log generator, real time process model discovery and process model improvement. Process model preparation aims to prepare the process model for the input of real time process log generator. Process model parsing applies reader and parser to retrieve a process model, either discovered from process mining or a priori process model. It aims to read and turn the process model into a required data structure for further processing. The real time process log generator utilizes complex event processing to generate events based on the given process model. Real time process model discovery is used to discover a process model based on the real time data generated by real time process log generator. Some data from the log generator will be stored in the

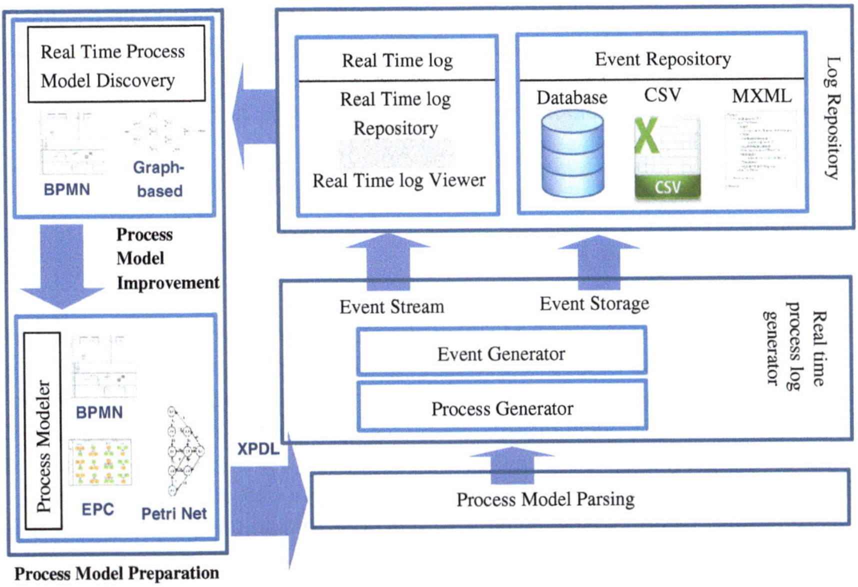

Fig. 1 Real time process log generator framework

repository, i.e., a database. Finally, process model improvement is a step to improve the discovered process model. This study will focus on the real time system instead of process model improvement. Each step will be explained in detail in the next section.

2.1 *Process Model Preparation*

Process model preparation requires a process model as the input for generating real time event data using real time process log generator. The process model can be obtained in some ways such as discovered using process mining and generated using some modeler tools. A discovered process model using process mining tool can be provided in various forms (e.g., Petri Net, BPMN, EPC, etc.). Due to various notations, workflow management coalition (WfMC) creates a standard language, called XML Process Definition Language (XPDL), for interpretability of various BP [13]. In this study, we apply XPDL by using the general definition of a process model based on BPMN.

A process model (P) consists of a set of activity (A), gateway (G), resource (R) and time information (TI). An activity is regarded as the step of process from the beginning until the end. The beginning of a process has a start event and the end of the process has an end event. Gateway controls the route of activity whether it goes to a sequence, parallel or selection gate. Most of the process model acknowledge the two prominent gateways; AND (parallel) and XOR (exclusive OR). Resource information provides resource properties in each activity. In real manufacturing environment, a resource can be considered as either a machine or a worker. To generate a random event, the process model has time information as designated by the users.

2.2 Process Model Parsing

Process model parsing is a layer to receive input, i.e., process model, for generating processes and events. To understand the procedure of process model parsing, it is necessary to acknowledge about process model standard language. The term process model (P) as aforementioned is used. Due to various workflow products, i.e., different modeling tools and management suites, XPDL is used in this study. It defines an XML schema for specifying the declarative part of business process.

The process model parsing consists of functions to read the XPDL file and map the contained elements into structured data objects. The read function is necessary to analyze the process flow since existing modeler tools on drawing BPMN can result a different format of XPDL. Thus, process model parsing reads the relevant attributes, i.e., activity, gateway, resources, and time. These elements are stored in the relevant data structure for log generator.

2.3 Real Time Process Log Generator (RT-PLG)

RT-PLG aims to execute a given process model and result an event log in real time manner. Since there might be complex events in the running application, it requires a specific engine for the execution. There are many complex event processing (CEP) engines being used worldwide. CEP engine will execute an input which contains various attributes and result a complex event according to the rule engine. Most of CEP engines have similar application framework. However, each engine has distinct features in regard to the vendors.

This study attempts to extend several layers according to the needs of process mining. There are two layers that extend the current CEP; process generator, and event generator. Process generator aims to create a random process in regard to designated arrival time. Once a process instance is created, it stimulates event generator to create events based on the predefined process model. Each of the layers will be explained in the next section.

2.3.1 Process Generator

Process generator receives an input from process model parsing. The input is to generate a process instance, called as a case.

Definition 1 *Case (C)* The notion of C is a set of event (E) for its one instance where a collection of all cases is an event log (L).

A case is created according to random generated number, defined by the user. Each case will have a particular flow, which is configured via the XPDL file. A flow contains a sequence of activities, where each activity may be handled by one or more machines; depending on the configuration file. This flow shall pinpoint what activity of a case shall go through in order to complete its flow.

Figure 2 depicts the execution flow of the proposed system in this work, and could be organized as follows:

1. Process log generator generates instance of process
2. Activity module ask resource module to use a particular resource
3. Resource module check the availability of the designated resource and return signal to activity module
4. Activity module shall start the activity if the designated resource is available, or put the product in the queue until available machine arise (triggered by a complete event of the required machine).
5. Activity module may produce three types of event:

   (i) Start: when a particular activity is started on the designated resource
   (ii) Wait: when the designated resource candidates are all being occupied
   (iii) Complete: when a particular activity is completed on the designated resource

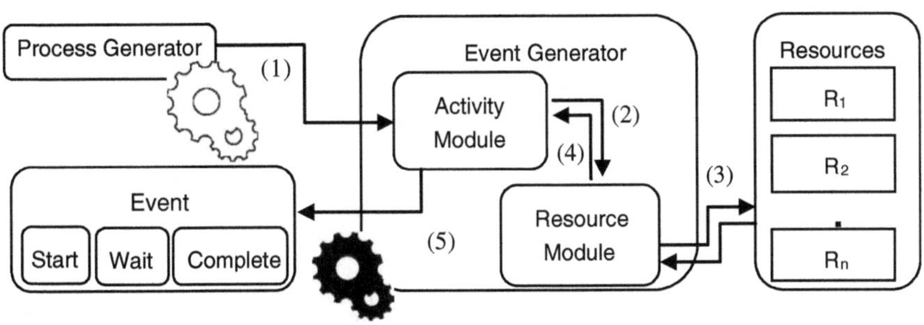

Fig. 2 The execution flow of RT-PLG

2.3.2 Event Generator

Each event in a case is generated in the event generator. The event is defined as follow.

Definition 2 *Event (E)* $E = A \times Y \times R \times T$ is a set of events, where A is a set of activities, Y is a set of event type (start, wait, complete), R is a set of resources and T is a set of timestamps.

The event generator has several modules such as activity and resource. The activity module produces an event related to activity property. In this module, it presents some general properties such as activity name followed by the case number. The resource module manages the resources which shall be used by a specific activity. If a case '5', at the front position of a queue, shall go through activity 'A' which needs to use either resource 'R1' or 'R2' to complete, resource module must check the availability of either resource 'R1' and 'R2' then assign the available one to case '5'. At this situation, a "Start" event will be raised indicating the beginning of activity 'A' on case '5'. Otherwise, case '5' shall wait in the case queue until one alternative resource, 'R1' or 'R2', is not occupied any longer. In this case, a "Wait" event shall be raised; indicating case '5' is waiting for an available machine to perform the next activity. When either resource 'R1' or 'R2' completes its current task, a "Complete" event indicating a finished activity on a particular resource shall be raised, calling up the next case in the queue, i.e., case '5'.

The events generated by this event generator later will be processed into a format to be used later in process mining. Similar to simulation tool, generated instance is instantiated in corresponding to specific arrival time. Assuming the products have been generated and the cases are already instantiated according to their respective process model configuration; the system will process each case and generate its events along the flow in a manner as described in Fig. 3.

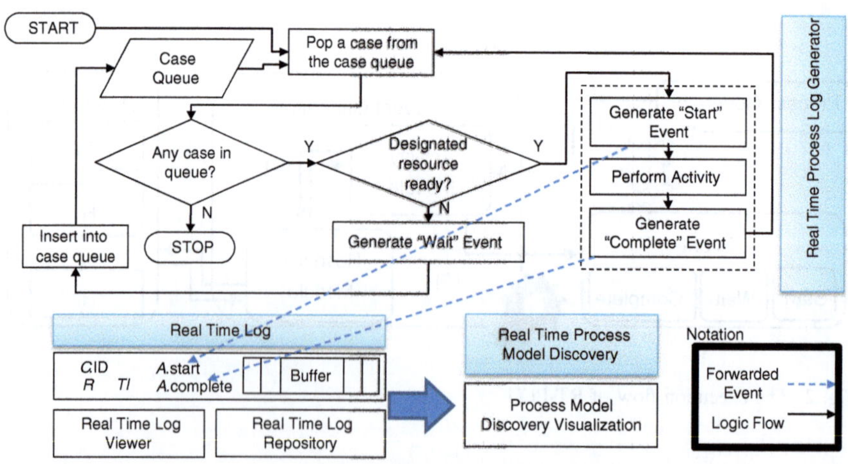

Fig. 3 Logic of the proposed framework and real time process model discovery

2.4 Real Time Process Model Discovery

Real time process model discovery aims to display the respective process from real time log. It should be noted that the result of process and event generator comprises of two; real time log and event log. Real time log consists of data updated by RT-PLG in the real time manner. Event log is regarded as stored or historical data for further analysis. This study will use real time log, instead of event log, to show the advantage of RT-PLG for real time process model discovery. Real time log stores each causal dependency information rather than all events. Hence, causal dependency and a real time log are defined as follows.

Definition 3 *Causal Dependency* Causal dependency is the set of two events which the earlier comes before the latter. Suppose there are two events, e_i and e_j. Two events which have causal dependency mean that e_j comes as soon as e_i finish.

Definition 4 *Real Time Log* (*L*) A real time log is a collection of causal dependencies. Causal dependency follows a trace, denoted as $\sigma \in E*$. $C = E*$ is the set of possible event sequences (traces describing a case). Then, $L \in \mathcal{B}(C)$ is a real time log. Note that $\mathcal{B}(C)$ is the set of all bags (multi-set) over C. Each element of L denotes a case.

In real world applications, real time log comprises of many attributes of processes. Since this study focuses on the usage of RT-PLG, this study limits the incoming events by considering only attributes for process mining, i.e., Case ID (C. ID), the start event of an activity (A. start), the complete event of an activity (A. complete), the resource (R) used on the related activity and the timestamp (TI). These events are stored in a specific buffer for further process.

The real time log is shown using real time log viewer and stored in the real time log repository in the specific format and updated in real time manner. Since the data is generated with an intense velocity and volume, it is obvious that a particular approach for handling the big data is necessary. This study limits the discussion of the big data for real time process model discovery and focuses on the potential impact of RT-PLG for the streaming process mining research.

The real time log in this study takes advantages on the traces of events which are based on timestamp. The traces of events contain sequence and can be seen as correlations as well as dependencies among events. Since this study attempts to show the potential of RT-PLG for real time process mining, real time process model discovery utilizes the general causal dependency on which the process model generated is based on the sequential events instead of finding other control flow (AND and XOR).

For the purpose of verification, the real time log is stored in the database. The stored data, later called as event log, contains a lot of information. Relevant information which is defined in Definition 2 is used. The database stored the actual information such as activity name, related resource which executes the activity, event type, and timestamp of the executed event. When the simulator reaches the end of designated events, there is a function to convert the data into a formal form such as MXML and XES format; exported in one single file.

3 Implementation Results

This section describes the implementation of the proposed framework. It explains the proposed framework scenario implementation including the verification of the logs produced.

The starting point of the framework is a process model. The process model is created using a process model notation, called business process modeling notation (BPMN). BPMN can be converted into a standard construct language, called XPDL. Hence, the notation, i.e., BPMN, is mapped into XPDL and then retrieved by the process model parsing. The process model parsing will send the event signal into CEP engine. In this study, Esper[1] is used for the CEP Engine. Afterward, the log repository will show the data according to user's selection. It means the real time data will be displayed in a web page and simultaneously stored in a repository. User may disable the display of real time data when only event logs are needed. At the final stage, the verification method using process mining tool, which is DISCO, is discussed.

To show the application of the proposed framework, a case study based on generated events from a process model is constructed[2] (Fig. 4). The process model, that has been redrawn using BPMN, is a manufacturing process which consists of 16 activities. Each activity has more than one resource which is predefined by the users in the process model. The activity time is also predefined by the users, e.g., using triangular distribution. To draw the process model, an existing process model editor is used. The process model editor enables the BPMN file to be exported into an XPDL format.

The XPDL is sent into the process model parsing. Process model parsing reads the XPDL and converts it into a certain format for generating complex events in the extended CEP engine. Esper, as one of the existing CEP engine, is used to generate a complex event based on the process model defined. Esper has been developed to enable rapid development of applications that process large volumes of incoming messages or events, regardless of whether incoming messages are historical or real-time in nature. Hence, Esper is used to execute an input from process model parsing.

Process generator and event generator produce cases and events respectively. The generated cases and events are stored as real time log in the format of JavaScript Object Notation (JSON) and used to discover the process model in a real time manner. This study runs the RT-PLG for 60 min and results the real time

[1]http://esper.codehaus.org/.

[2]https://www.promodel.com/Products/ProcessSimulator.

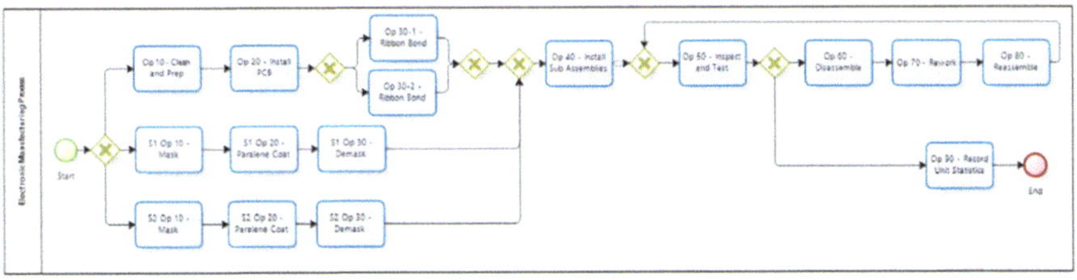

Fig. 4 An example of process model for experiment

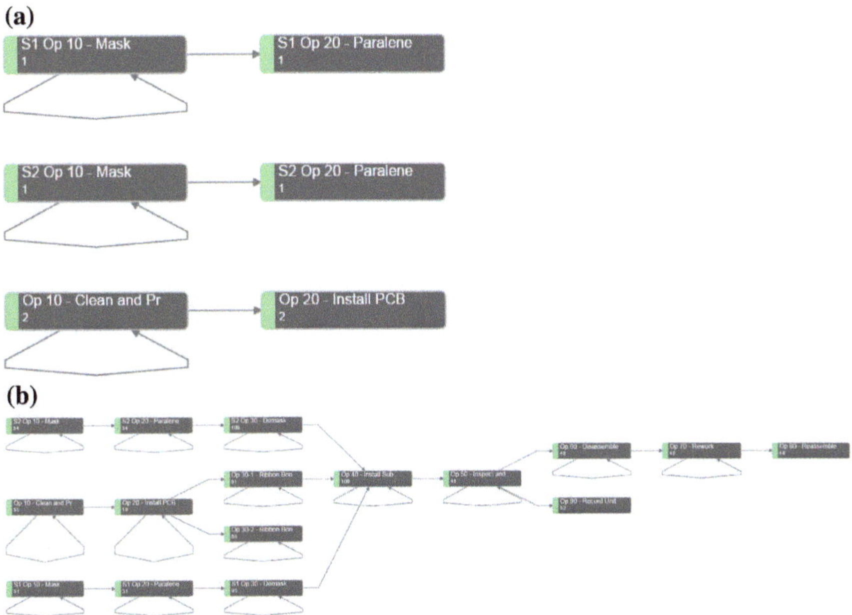

Fig. 5 **a** Discovered process at time t, **b** Discovered process in real time after time t

process discovery as shown at Fig. 5. Figure 5a shows the discovered process at a certain time t. The ongoing process model when the generated events of some cases are reaching end nodes is shown at Fig. 5b. Since the real time process discovery uses causal dependency, the discovered process is in a sequence literally. A self-directed arrow on some particular nodes represents the processing time required by a task to complete.

A process mining tool should be used to verify the produced event logs. The event logs, in the format of MXML, are used for the verification method. In this study, a commercial process mining tool, DISCO,[3] is utilized. The procedure to verify the event logs is as follows. First, the event log produced by RT-PLG is

[3]http://fluxicon.com/disco.

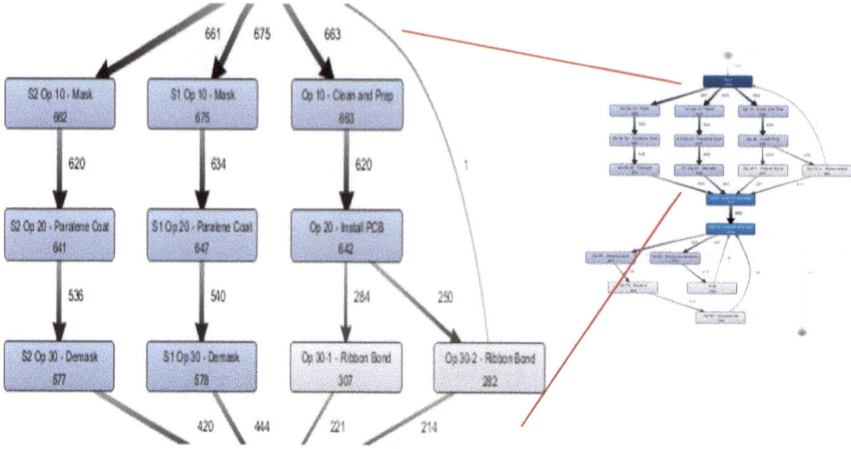

Fig. 6 Verification result using DISCO

stored into a repository with a particular format for the input of process mining tool, i.e., DISCO. Since the data is obtained in a real-time manner, a premature system termination will cause a particular case fails to reach the end of process and produce an incomplete event log. The incomplete data is pruned and the filtered data is used for discovering the process model. For the verification, 2000 cases and 11,836 events were generated. Figure 6 shows the discovered process model of the produced event logs using DISCO.

4 Conclusion

This paper presented a framework for real time process log generator. The real time process log generator applied CEP which utilized process model as the input. Process model can be either an a priori model created by a user or a discovered process model using process mining. The process model, before processed in the CEP engine, was retrieved using process model parsing. The output of process model parsing was used in the CEP engine, which executes two layers; process and event generator. At last, the produced event log is immediately used to discover the process model rather than stored in a log repository.

The work presented in this paper is considered as an extension of the previous work to address the problem of random generation of business process in real time manner [8, 9, 12]. This work differs from the previous work [14] on which it attempted to deal with real time log without any storage. Real time process log generator has several advantages due to the rising of streaming process mining approach. First, the rising of process mining for real world application needs more reliable event log. There is limited knowledge regarding a tool that can support, in particular, streaming process mining algorithm. Second, it is necessary to support process enhancement using both animation and simulation. Animation is a way to replay the process based on event log. On the other hand, simulation is an operation to either analyze the enhanced process model or roughly monitor the running imitation process in comparison to real world process. This can also be a supporting tool for detecting a concept drift in process discovery. Thus, the future work of this study will be toward establishing a life cycle procedure to support both process enhancement and monitoring simultaneously.

Some issues are open for further work. Concerning a complex process, it is possible to characterize the semantic of the XPDL generated by various tools. The consistency and homogeneity of the XPDL is also another potential research work. Due to various process modeler tools, it is necessary to integrate and build a common construct for any business process. Hence, there are empty rooms to manipulate such tool by constructing a general business process language which conforms to data from various application domains.

References

1. Aalst, W.M.P.: Process Mining: Discovery, Conformance and Enhancement of Business Processes. Springer, Berlin (2011)
2. Aalst, W.M.P., Reijers, H.A., Weijters, A.J.M.M., van Dongen, B.F., Aalves de Mediros, A. K., Song, M.S., Verbeek, H.M.W.: Business process mining: an industrial application. Inf. Syst. **32**, 713–732 (2007)
3. Aalst, W.M.P., Weijters, A., Maruster, L.: Workflow mining: discovering process models from Event Logs. IEEE Trans. Knowl. Data Eng. **16**(9), 1128–1142 (2004)
4. Burattin, A., Sperduti, A., Aalst, W.M.P.: Heuristics Miners for Streaming Event Data. http://arxiv.org/pdf/1212.6383v1.pdf (2012)
5. Zelst, S.J., Burattin, A., Dongen, B.F., Verbeek, H.M.W.: Data streams in ProM 6: a single-node architecture. http://ceur-ws.org/Vol-1295/paper6.pdf (2014)
6. Maggi, F.M., Burattin, A., Cimitile, M., Sperduti, A.: Online process discovery to detect concept drifts in LTL-based declarative process models. In: OTM 2013, LNCS 8185, pp. 94–111 (2013)
7. Maggi, F.M., Burattin, A., Cimitile, M.: Lights, camera, action! Business process movies for online process discovery. Workshop on Theory and Applications of Process Visualization (2014)
8. Burattin, A., Sperduti, A.: PLG: a framework for the generation of business process model and their execution logs. In: Proceedings of the 6th International Workshop on Business Process Intelligence 2010. Stevens Institute of Technology; Hoboken, New Jersey, USA. doi:10.1007/978-3-642-20511-8_20 (2010) 13 Sept 2010
9. Vanden Broucke, S.K.L.M., Vanthienen, J., Baesens, B.: Straightforward petri net-based event log generation in ProM (2014)
10. Banks, J., Carson, J., Nelson, B., Nicol, D.: Discrete-Event System Simulation. Prentice Hall, Upper Saddle River (2001)
11. Borschchev, A.: The big book of simulation modeling. Multimethod modeling with AnyLogic 6. AnyLogic North America (2013)
12. Online Business Process Software. https://www.bpsimulator.com/

13. WFMC.: Workflow management coalition workflow standard: workflow process definition interface—XML process definition language (XPDL) (WFMCTC-1025). Technical report, WMC, Lighthouse Point, Florida, USA (2002)
14. Oliveira, J.J.: Business process discovery in real time. https://fenix.tecnico.ulisboa.pt/downloadFile/395142792129/Extended%20Abstract.pdf

Medical Data Unification Using Ontology-Based Semantic Model Structural Analysis

Mariusz Chmielewski and Piotr Stąpor

Abstract In the paper we will present a cohesive view on medical data integration and migration mechanisms for providing and maintaining a medical system interoperability. It is well known, that despite very well defined semantics, the medical domain utilizes many data formats, and various data records along with domain values (datatypes and enumerations). Therefore, in healthcare systems, the integration itself becomes an important and crucial issue. To solve such problems, a variety of ETL tools can be used. Our method and tools provide several features, which extend the capabilities of such mechanisms by delivering semantic analysis of data. Utilized ontology-based features of reasoning mechanisms, can be used for identity and consistency validation. The method delivers also analytical approach based on ontology-structural analysis applied for ontology alignment process. This technique produces recommendations for a knowledge engineer by identifying ontology elements correspondences.

Keywords Medical ontologies · Ontology modelling · Data integration · Structural analysis

1 Introduction

The paper summarizes characteristics and features of semantic model structural analysis applied for ontology mapping techniques in the area of medical data unification. Our main assumption for formulating main research problem is that each healthcare related system operates on a specific set of databases (patients, diseases, diagnostics techniques etc.). If medical data is stored in a traditional

M. Chmielewski (✉) · P. Stąpor
Cybernetics Faculty, Military University of Technology,
Gen. Kaliskiego 2, 00-908 Warsaw, Poland
e-mail: mariusz.chmielewski@wat.edu.pl

P. Stąpor
e-mail: piotr.stapor@wat.edu.pl

© Springer International Publishing Switzerland 2016

J. Świątek et al. (eds.), *Information Systems Architecture and Technology: Proceedings of 36th International Conference on Information Systems Architecture and Technology – ISAT 2015 – Part III*,
Advances in Intelligent Systems and Computing 431, DOI 10.1007/978-3-319-28564-1_12

database our approach may still apply as many attempts of automatic database to ontology transformations have been already conducted and recognized [1]. In this paper we will define semantics through the formulation of an ontology on which we will create instance base, thus separating terminology and instantiated data. Medical domain has been thoroughly researched and described in the context of ontology modelling and well-structured knowledge and semantics. The use of ontology modelling in a discussed domain is focused on structuring and organizing terminologies within e.g. medical treatment, medical diagnostics, diseases and organisms taxonomies. In this work we will be focusing only on these medical semantic models, which can be transformed into compatible Description Logic [2] statements and OWL representations. Formulated structural analysis method uses formal ontology models and transforms them into multilayer weighted graph structures. The advantages of such structures come from the availability of analytical algorithms and quantitative approach to node and edge importance analysis. These factors can be successfully used as a structural similarity measures, which in composition with string (label) similarity matching measures, deliver accurate multi-criteria matching methods. Having well-defined quantitative approach, presented method evaluates the similarity of elements (mainly concepts and individuals) within a pair of aligned ontologies. The method assumes an interaction with a knowledge engineer, who verifies and validates proposed semantic mappings. Built-in capabilities of OWL and RDFS languages simplify the mapping of an ontology element. The results of proposed method can be observed after application of DL reasoning mechanisms. Instances expressed in specific terminologies are mapped onto each other and classified according to developed ontology alignment. Description of the semantic unification method will be followed by an example and its analysis demonstrating the method's capabilities and drawbacks.

2 Ontology Quantitative Analysis Using Structural Approach

The structural approach, presented in this article, is derived from a graph theory and utilizes weighted graph structures as representations of both the ontology and the instance base. The method implies that semantic models can be mapped into multigraphs with vertices being either concepts or individuals and edges representing particular binary relations [3]. Ontology transformation process results in a set of weighted-graph structures used as homogeneous layers—carrying similar relation type. Therefore, the layer is considered a subgraph used to emphasize the specific aspect of semantics in the domain. For example a concept generalization layer, called Concept Taxonomy Graph, can be examined for distribution of concepts within given ontology hierarchy, semantic capacity and so on. Structural analysis performed on the set of extracted layers enables us to capture characteristics of "importance" for given concept within the ontology. On the basis of evaluated measures, presented method solves an assignment problem—matching concepts between given pair of ontologies. As a result we will obtain a mapping scheme for two terminological systems—thus solving the problem of integrating the terminologies. Usage of the structural analysis give us another advantage—an opportunity to apply quantitative approach using node and edge centrality measures. We apply them as importance evaluators for concepts and relations. Obviously the importance of a given element is subjective, depending on the defined analysis criterion. That is why the proposed set of measures has been examined and analyzed in terms of given element usage in model definitions, its centrality in the structure and authority. The structural method has been thoroughly described in [3]. Note that measures evaluation may generate new knowledge, therefore we can consider it as a reasoning process. For example CNC measure (also called link relevance [4]) can be used to detect a potential (hidden) linkage between various pairs of concepts [3]. Modifications of that measure help also to evaluate the strength of that hidden correspondence. Below are described only the selected aggregated graph measures, that have been used in this article for the presented method portrayal. The first group of measures constitutes of taxonomical measures, for both concepts (CCTM—Concept Capacity Taxonomy Measure) and properties (SRCTM—Structural Relationship Capacity Taxonomy Measure). CCTM may deliver information concerning given concept's definition specificity and accuracy (Fig. 1).

The measure may also be used to evaluate the taxonomical subgraph's importance in which a certain concept exists. Reference [3] introduces the term semantic capacity, which reflects the given concept's specialization and its relative position in a taxonomy subgraph. The most specific concepts—taxonomy tree leafs—get maximum value, while the most general ($\top$) receive minimum value. The richness is

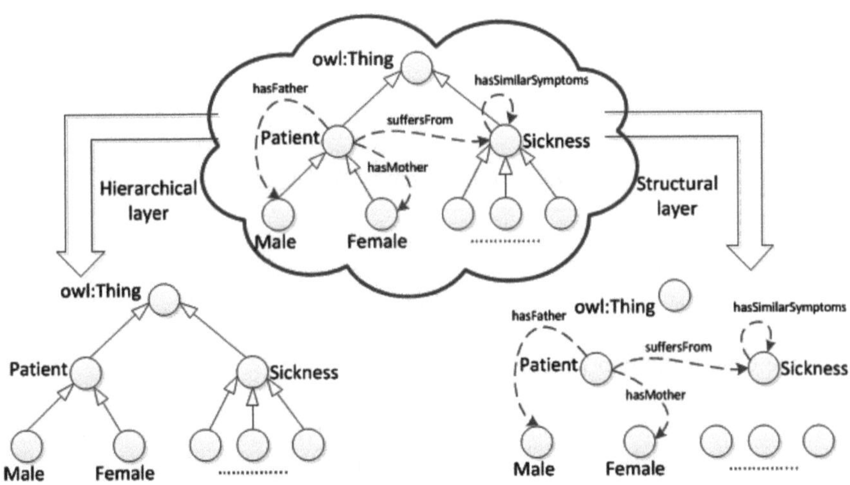

Fig. 1 Hierarchical and structural layers of an ontology

assessed as the number of inheritance levels has an influence over the resulting value. CCTM reflects concept's multiple inheritance abilities by taking into account the length of the longest generalization path in which the given concept exists. The measure provides additional profit, as it can be utilized for observing taxonomy changes incorporated by the reasoning engine's reclassification algorithm [3]. The ACCTM (Absolute Concept Capacity Taxonomy Measure) and ASRCTM (Absolute Structural Relationship Capacity Taxonomy Measure) are derivatives of the CCTM and SRCTM. Their values can be interpreted as the semantic capacity evaluated with the respect to a deepest specialization path of an entire taxonomy [3].

$$\vec{m}_{ACCTM}(c_i) \triangleq \frac{len\left(path_H^{\max}(c_i, c_0)\right)}{\max\limits_{c \in C}\left(len\left(path_H^{\max}(c, c_0)\right)\right)} \tag{1}$$

$$\vec{m}_{CCTM}(c_i) \triangleq \frac{len\left(path_H^{\max}(c_i, c_0)\right)}{len\left(path_H^{\max}(c_i)\right)}, \tag{2}$$

where:

- $path_H^{\max}(c_i, c_0)$ is the longest path in taxonomy that connects concepts c_0 and c_i.
- $path_H^{\max}(c_i)$ is the longest path starting at c_0 that passes through c_i.
- Function *len* returns the path's length.

The CNC (Conjunctive Neighbors Coefficient) measure is derived from Jaccard index. It can be used for the purpose of finding the hidden connections between concepts, as well as for the removal of useless relations (low importance). CNC tests if two nonadjacent nodes can (or should) be associated with each other, according to the strength of their common neighborhood.

Let $N_V(v')$ be defined as a neighborhood of vertex v'.

$$N_V(v') \triangleq \{v \in V \backslash \{v'\} | (v, v') \in E \vee (v', v) \in E\} \tag{3}$$

The

$$N_V^{\cap}(v_i, v_j) \triangleq N_V(v_i) \cap N_V(v_j) \tag{4}$$

and

$$N_V^{\cup}(v_i, v_j) \triangleq N_V(v_i) \cup N_V(v_j) \tag{5}$$

form the CNC definition [3]:

$$\widehat{m}_{CNC}(v_i, v_j) \triangleq \frac{\left|N_V^{\cap}(v_i, v_j)\right|}{\left|N_V^{\cup}(v_i, v_j)\right|} \tag{6}$$

LNCC measure (Local Node Clustering Coefficient for Directed Unigraphs) [3, 12] is given as

$$\tilde{m}_{LNCC}(v_i) \triangleq \frac{E_i^{Neighbours}}{k_i(k_i - 1)} \tag{7}$$

for $k_i > 1$ and 0 otherwise, where $E_i^{Neighbours}$ is the number of connections between neighbors of v_i in a given directed skeleton, while k_i represent the number of v_i neighbors in a given directed skeleton. LNCC can be used to [3]:

- measure concept's importance in terms of existing structural relationships and as an element used to connect ontology modules,
- identify clusters of concepts (nodes) strongly semantically related with each other and indicate their (clusters) nature by evaluating the clustering formulation,
- filter and determine set of concepts falling into a very similar semantics for modularization purposes.

The normalized degree measure (CDTM) complements semantic capacity by providing additional information on sub-taxonomy richness (in-degree: CDTM−, out-degree: CDTM+) in terms of subsumption axiom definitions (inheritance multiplicity) [3]. The plot of the topological coefficients (TC) [5] can be used to estimate the tendency of the nodes in the network to have shared neighbors. SCBM (Structural Concept Betweenness Measure)—Let $s, v, t \in V$ and σ_{st} be the number of shortest paths from vertex s to t. Through $\sigma_{st}(v)$ we shall denote the number of shortest paths from vertex s to t containing vertex v within. The SCBM of an ontology representation multigraph is defined as [6]:

$$C_B(v) \triangleq \sum_{s \neq v \neq t} \frac{\sigma_{st}(v)}{\sigma_{st}} \tag{8}$$

According to [3], SCBM can be extremely useful for a task of graph connectivity evaluation. It can be utilized to identify the given node's importance in terms of its occurrence in the shortest paths. It does not concentrate on the length parameter but on the quantity of shortest paths [3]. The algorithm for SCBM evaluation has a high complexity, thus approximation should be used for a practical analysis of large ontologies. Lastly, the we shall introduce the SCCM (Structural Concepts Closeness Measure) measure. Its purpose is to rank the connectivity between concepts through the whole semantic model [3]. The designated value should be interpreted as an "average distance" of reaching a given concept through semantic paths originating from the other nodes in the structure [3]. For an ontology representing multigraph the SCCM

$$\vec{m}_{SCCM}(c_i) \triangleq \sum_{c_j \in C \setminus \{c_i\}} \left(\min_{path(c_i,c_j) \in Paths(c_i,c_j)} \{len(path(c_i,c_j))\} \right), \tag{9}$$

where:

- $Paths(c_i, c_j)$ is a set of all paths starting from c_i and ending at c_j.
- $path(c_i, c_j)$ is a single path with source c_i and destination c_j.

Many of recognized measures have been thoroughly examined on the ground of semantic networks, conceptual graphs and ontology structures. Research in this field provides the characteristics of aforementioned measures in the context of semantic model evaluation and the usefulness of these measures applied to assess the importance of ontology elements. (The element which is the closest to the other elements, most referenced element, the element which has the largest neighborhood etc.) Many of addressed measures evaluate the element neighborhood or its distance in the structure. Quantitative analysis on specific ontology structures is performed using aggregated measures (graph's nodes' and edges' importance measures) which emphasize various aspects of element's semantic importance.

Presented measures deliver evaluation capabilities of semantic model elements, but they are computational-intensive. Time-complexity evaluation for referenced set of measures has been presented in [3]. In many cases the evaluation process can be designed so that it shares calculations semi-products between evaluation algorithms which significantly speeds up the process. In some cases approximation algorithms can be used to estimate the value of a measure in a given, constrained time frame.

3 The Mapping Method

There are two widely recognized strategies of semantic unification: ontology merging and ontology mapping (alignment) [7, 8]. Existing publications present various methods for performing the task, using tailored algorithms but most of all presenting rules and guidelines. Presented research has been based on the ontology mapping as it is the most suitable and aligned to ETL requirements. Data migration requires not only technical links between data sources but most of all semantic unification (understanding) of transferred data records. In described research we will consider pairs of terminologies containing unique overlapping domain descriptions. By overlapping we acknowledge existence of similar or semantically equivalent concepts, roles or individuals. The method operates on the level of data sources thus it requires execution of two activities: Schema overview for terminology extraction—involving detailed analysis of tables, attributes, enumerated types and relations in order to select key concepts and capture binary structural

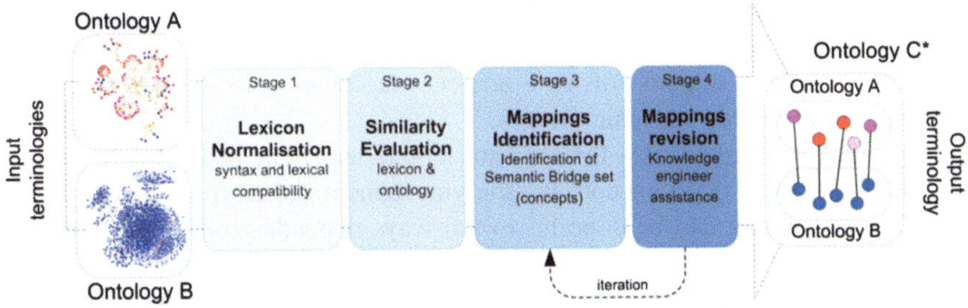

Fig. 2 The ontology mapping process diagram-main stages

relations. Having these elements, the transformation mechanisms can start the ontology development process involving also specific axiom formulations (simple ones). Data querying, transformation and loading—this process aims at elementary data operations and mainly involves transformation of data records into a set of ontology instances reflecting the source schema delivered in form of an ontology;

Extracted terminology should be subjected to a human refinement process ensuring modelling consistency and proper, required level of ontology entailments (axioms). The knowledge engineer should be supported by reasoning mechanisms, which should be responsible for automatic model validation in terms of concept and knowledge base satiability. Application of DL languages ensures the availability of reasoners and validation of developed terminologies and instance bases. That is why the application of ontologies is so valuable (Fig. 2).

The ontology mapping process is composed of four main stages. Lexicon normalization is the activity required for most cases of lexicons where we can suspect various forms of vocabulary used on ontology: syntax errors, misplaced characters, complex labels. Similarity evaluation is the evaluation of semantic model elements (on the level of terminology) using defined similarity and importance measures. The process is composed of two parts: the evaluation of lexicon elements using string distance measures [9] and the evaluation of ontology elements using structural importance measures applied on Concept Taxonomy Graph and Ontology Structure Multi-Graph [3]. The similarity measurement for pairs of concepts can be recognized on the lexicon level showing similar labels or phrases for given ontology element as well as on the axiom definition level (logical) demonstrating similar element definitions in terms of used DL language statements. Mapping identification is the most crucial part of the mapping procedure, involving the evaluation of multicriteria assignment problem. In this stage the preference model is built, which defines the priorities of the evaluated similarity measures and calculates the optimal concepts' assignment from A to B ontologies. Mappings revision is the last not automated stage of the mapping, requiring knowledge engineer assistance and interactions. Based on the previously obtained results, the algorithm presents the most similar concepts in the given pair of ontologies. Having this information the knowledge engineer verifies proposed assignment and decides how to implement the semantic bridge [1] for a given pair of concepts. In result a new ontology is created containing imports of two source ontologies and elements mapped according to identified semantic bridges.

The mapping method uses the semantic bridge as a statement, which binds in some defined form two concepts by applying constructs in given ontology language. Such binding can be done in several ways depending on the mapped element. The most interesting and valuable is concept mapping. Utilizing DL statements we can formulate following mappings: concept equivalency mapping—definition of concept through the application of CEA (Concept Equivalence Axiom); concept inclusion mapping—definition of concept through the application of CIA (Concept Inclusion Axiom); value restriction mapping—definition of a mapping through the usage of value restrictions—existential restriction and universal value restriction; cardinality restriction mapping—definition of a mapping through the usage of qualified cardinality restriction.

A semantic bridge is a set of axioms, which defines complex relations or statements between concepts taken from source ontologies O_1, O_2, O_3. C_1, C_2, C_3 and R_1, R_2, R_3 are concepts and relations between them taken from mentioned ontologies, such that C_i, R_i are elements of O_i.

The mapping of ontology elements presented in the Table 1 specifies two groups of available modelling constructs. First group of semantic bridges can be built on the basis of CEA and CIA axioms which formulate the equivalence and inclusion between pairs of concepts. The second group emphasize complex relationships formulated as value or cardinality restrictions, that is axioms formulated on the basis of quantification. The statements presented in the Table 1 demonstrate only possible forms of such axioms. The main requirement for such constructs is to relate two concepts through a restriction e.g. $C_3 \equiv (\exists R_1 . C_1)$.

Given previous statement we can identify that there exists a complex relation between C_3 and C_1 as C_3 defines a set of individuals, which possess at least one relation R_1 connecting them to individuals from (restricted to) C_1.

In order to construct fully covered ontology alignment the presented mapping method may require extensions of input ontologies. To extend the model we can utilize one of three verified strategies. (Strategy 1) Addition of a new ontology

Table 1 Formal specifications of various types of semantic bridges used in mapping procedures

Type of mapping	Formal specification of semantic bridge
Concept equivalency mapping	$((C_1 \equiv C_3) \wedge (C_2 \equiv C_3)) \Rightarrow (C_1 \equiv C_2)$
Concept inclusion mapping	$((C_1 \sqsubseteq C_3) \wedge (C_2 \sqsubseteq C_3)) \Rightarrow ((C_1 \sqsubseteq C_2) \vee (C_2 \sqsubseteq C_1))$
Value restriction mapping	$C_3 \begin{smallmatrix}\equiv \\ \sqsubseteq\end{smallmatrix} \left(\begin{smallmatrix}\leq \\ \geq\end{smallmatrix} n\, R_1 \cdot C_1 \right) \begin{smallmatrix}\sqcap \\ \sqcup\end{smallmatrix} \left(\begin{smallmatrix}\leq \\ \geq\end{smallmatrix} m\, R_2 \cdot C_2 \right)$
Cardinality restriction mapping	$C_3 \begin{smallmatrix}\equiv \\ \sqsubseteq\end{smallmatrix} \left(\begin{smallmatrix}\leq \\ \geq\end{smallmatrix} n\, R_1 \cdot C_1 \right) \begin{smallmatrix}\sqcap \\ \sqcup\end{smallmatrix} \left(\begin{smallmatrix}\leq \\ \geq\end{smallmatrix} m\, R_2 \cdot C_2 \right)$

module is done by using the merging scheme. This permits to produce one merged model, with separate name spaces while preserving the model logical consistency. It is required to check if the concepts placed in separate modules are not redundant. (Strategy 2) Introducing new ontology elements should be done with the use of DL axioms to bind the concepts with other parts of ontology. If this cannot be achieved due to the lack of available semantics, new branch of concepts should be introduced with additional domain closure axioms owl:disjointWith. (Strategy 3) The usage of OWL language mapping capabilities strategy is suitable for integrating already developed models and assumes that two separate ontology models can be combined in the form of previously described semantic bridges set. Extensions of the model should be divided into the extension of terminology (TBox) (concepts, roles) and the instance base (ABox). In the first case, the extension concentrates on concepts and relations, in the second one—on the instance placement in a knowledge base. However, instances are often omitted in terms of terminology definitions (with the exception of the enumerated concept definitions).

4 Experimental Evaluation—the OBO Ontologies

The purpose of ontology modelling is the delivery of well-defined, logically consistent conceptual model. Unfortunately there is no strict set of common guidelines and rules for ontology model construction therefore their complexity may vary. In consequence ontologies may differ in several aspects: (a) used formalism and modelling approach—(taxonomy, frame, axioms, conceptual graphs) (b) modelling purpose—(knowledge retrieval and sharing, natural language processing, knowledge organization, problem solving, classification, etc.) (c) construction approach—(top-down, bottom-up, merging, mapping, etc.) (d) domain coverage, model implementation, size, precision, complexity, etc. Analyzed further ontologies have been chosen to comply with similar model complexity and size. Existing medical terminologies in many cases semantically overlap each other in two different ways. First case refers to the situation in which the similarity originates from concepts' labels existing on the level of lexicons, however using different definitions-axioms. In other words the same concepts (meaning the same concept labels—phrases) may have different logical definitions. For demonstration of structural analysis capabilities, we have chosen a set of compact ontologies describing chemical domain closely related to analyzed medical descriptions. Problems that we had to face were the ones connected with the identification of ontologies which: are semantically related and share similar vocabularies, are saturated with structural relationships creating ontology structures with significant graph diameter, have complex definitions of concept and structural relation taxonomies emphasizing semantic capacity of formulated terminology, have rich set of axioms in concept definitions.

Table 2 Overall comparison of ontologies

	Lipid ontology	Chemical-v2
DL dialect	$\mathcal{ALCHIN}$	$\mathcal{SHOIF}^{(D)}$
Total concepts	715	87
Primitive	447	83
Defined	268	4
Total properties	46	7

In order to present most significant and reasonable measures' values (which clearly demonstrate structural analysis usage in terms of semantic model importance measurement) the obtained evaluation results have been filtered for reading convenience.

The OBO (The Open Biological and Biomedical Ontologies) project is a collaborative effort to gather a reliable knowledge from a field of biomedicine in a form of ontologies [10]. A compact obo ontology—Lipid Ontology—has been analyzed using proprietary software—Terrana [3, 11] (Table 2).

The lack of any LNCC magnitude demonstrates the sparsity of ontology structure. The SCBM non-zero values indicates the prominent importance of the "Lipid" concept. Based on obtained values and ontology visualizations it can be concluded that "Lipid" is the heart of this ontology structure. The topological coefficient shows high values for vertices which mark out as having large number of shared neighbors. As presented, the structural approach can be successfully used for semantic models evaluation, especially the large ones, which are too complex to be visualized. The results (for presentation purposes) have been sorted with a respect to chosen set of significant structural measures and aggregated measures (descending order), which are obtained for a given concept by summing all of its scores. The similar structural analysis has been done for another ontology—"chemistry-v2".

5 Conclusion

Presented method extends available ETL approaches and moves the data transformation mechanisms from tool-oriented programming domain, into semantic modelling domain. In our opinion, operating on the level of data semantics is more convenient, especially while designing the ETL process and resolving complex mapping schemes.

Many of available ETL tools and environments deliver simple data transformation utilities relaying on the operator's skills and data understanding. Application of designed automated mapping method resolves that issue. On one hand method delivers an automatically verifiable semantic representations through the application of DL language and reasoning facilities, and on the other one—the mapping recommendations which utilize structural approach and string similarity measurement (Table 3).

Table 3 Values for selected concepts of Lipid ontology

RDFResource	PageRank	Authority	Hubs	CDTM	CDTM-	CDTM+	SCCM	TC	CCTM	ACCTM	SCBM
Lipid	0,0014	0,0033	1,0000	1,0000	1,0000	1,0000	0,0500	0,0000	0,5333	0,5333	1,0000
Acyl Ester Chain	0,0014	0,1890	0,0008	0,0476	0,0500	0,0455	0,0256	1,0000	1,0000	0,6000	0,0000
Vinyl Ether Chain	0,0063	0,1890	0,0008	0,0476	0,0500	0,0455	0,0256	1,0000	1,0000	0,5333	0,0000
Ether	0,0014	0,1890	0,0008	0,0476	0,0500	0,0455	0,0256	1,0000	1,0000	0,5333	0,0000
Prenyl	0,0014	0,1890	0,0008	0,0476	0,0500	0,0455	0,0256	1,0000	1,0000	0,4667	0,0000
Acyl Chain	0,0014	0,1890	0,0008	0,0476	0,0500	0,0455	0,0256	1,0000	0,8889	0,5333	0,0000
Phytyl	0,0014	0,1890	0,0002	0,0238	0,0500	0,0000	0,0000	1,0000	1,0000	0,4667	0,0000
Sphingoid Base Chain Group	0,0014	0,1890	0,0008	0,0476	0,0500	0,0455	0,0256	1,0000	0,8750	0,4667	0,0000
Carbon Chain Group	0,0014	0,1890	0,0008	0,0476	0,0500	0,0455	0,0256	1,0000	0,6364	0,4667	0,0000
LC Methoxy mycolic acid	0,0014	0,0001	0,0002	0,0000	0,0000	0,0000	0,0000	0,0000	1,0000	1,0000	0,0000
Organic Group or Ring System	0,0014	0,5668	0,0014	0,1190	0,1500	0,0909	0,0256	1,0000	0,0000	0,0000	0,0000
LC Alpha prime mycolic acids	0,0014	0,0001	0,0002	0,0000	0,0000	0,0000	0,0000	0,0000	1,0000	0,9333	0,0000
Ubiquinone ring	0,0015	0,0001	0,0002	0,0000	0,0000	0,0000	0,0000	0,0000	1,0000	0,6000	0,0000
Alcohol	0,0014	0,0001	0,0002	0,0000	0,0000	0,0000	0,0000	0,0000	1,0000	0,6000	0,0000
Aldehyde	0,0014	0,0001	0,0002	0,0000	0,0000	0,0000	0,0000	0,0000	1,0000	0,6000	0,0000
Biological Entity	0,0014	0,0001	0,0002	0,0000	0,0000	0,0000	0,0000	0,0000	0,3333	0,3333	0,0000
Entity	0,0014	0,0001	0,0002	0,0952	0,2000	0,1818	0,0000	0,0000	0,0667	0,0667	0,0000
Polyatomic Entity	0,0014	0,0001	0,0002	0,0000	0,0000	0,0000	0,0000	0,0000	0,2667	0,2667	0,0000
Independent Continuant Entity	0,0014	0,0001	0,0002	0,0000	0,0000	0,0000	0,0000	0,0000	0,2000	0,2000	0,0000
Continuant_Entity	0,0014	0,0001	0,0002	0,0000	0,0000	0,0000	0,0000	0,0000	0,1333	0,1333	0,0000
owl:Thing	0,0014	0,0001	0,0002	0,0000	0,0000	0,0000	0,0000	0,0000	0,0000	0,0000	0,0000

The application of mapping schemes in form of semantic bridges delivers an automatic data transformation from one terminology into the other. The concept of automatic data transformation utilizes features of semantic web languages and DL reasoning process results. The method has its drawbacks related mainly to tasks of measuring and developing concept alignments. Solving the multi-criteria assignment problem, while considering measures evaluated for a lexicon (concepts/individuals/relations naming) and ontology characteristics, results in high algorithm time-complexity. However this task is not frequently executed, as we require only evaluation of measures for terminology and this part of semantic model is rarely changing. Applied structural analysis requires some characteristics of analyzed models, and unfortunately majority of available medical ontologies lack some crucial characteristics connected with structural relationship saturation. The method itself can be adapted to other ways of evaluating an ontology alignment—focusing rather on lexicon levels and strings than on the graph-based or axiom based similarity measures. Our further research will be aimed at extending the set of semantic similarity measures for evaluating the axiomatic layer in ontologies. The task can be proven to be complicated as it depends on the modelling language and forms of defined axioms and even unification of medical data itself is a very challenging problem (Table 4).

Table 4 Values for selected concepts of chemistry-v2 ontology

RDFResource	CCTM	ACCTM	PageRank	Authority	Hubs	CDTM	CDTM-	CDTM+	Closeness Centrality	SCCM
BiologicalProcess	0,5000	0,1429	0,0113	0,0074	0,9976	0,7500	0,0000	0,7500	0,5000	1,0000
PhysicalMaterialEntity	0,1429	0,1429	0,0113	0,0074	0,0074	1,0000	1,0000	1,0000	0,0000	0,0000
Molecule	0,2857	0,2857	0,0133	0,9976	0,0074	0,7500	0,7500	0,0000	0,0000	0,0000
Aldohexose	1,0000	1,0000	0,0113	0,0074	0,0074	0,0000	0,0000	0,0000	0,0000	0,0000
Al_3plus	1,0000	0,7143	0,0113	0,0074	0,0074	0,0000	0,0000	0,0000	0,0000	0,0000
Be_2plus	1,0000	0,7143	0,0113	0,0074	0,0074	0,0000	0,0000	0,0000	0,0000	0,0000
Aldose	0,8571	0,8571	0,0113	0,0074	0,0074	0,0000	0,0000	0,0000	0,0000	0,0000
Hexose	0,8571	0,8571	0,0113	0,0074	0,0074	0,0000	0,0000	0,0000	0,0000	0,0000
Ar	1,0000	0,5714	0,0113	0,0074	0,0074	0,0000	0,0000	0,0000	0,0000	0,0000
C	1,0000	0,5714	0,0113	0,0074	0,0074	0,0000	0,0000	0,0000	0,0000	0,0000
HydrogenChloride	1,0000	0,5714	0,0113	0,0074	0,0074	0,0000	0,0000	0,0000	0,0000	0,0000
Lipid	1,0000	0,5714	0,0113	0,0074	0,0074	0,0000	0,0000	0,0000	0,0000	0,0000
Water	1,0000	0,5714	0,0113	0,0074	0,0074	0,0000	0,0000	0,0000	0,0000	0,0000
Actinide	1,0000	0,4286	0,0113	0,0074	0,0074	0,0000	0,0000	0,0000	0,0000	0,0000
O	0,8000	0,5714	0,0113	0,0074	0,0074	0,0000	0,0000	0,0000	0,0000	0,0000
MetabolicProcess	1,0000	0,2857	0,0113	0,0074	0,0074	0,0000	0,0000	0,0000	0,0000	0,0000
MonoatomicIon	1,0000	0,2857	0,0113	0,0074	0,0074	0,0000	0,0000	0,0000	0,0000	0,0000
Metal	0,7500	0,4286	0,0113	0,0074	0,0074	0,0000	0,0000	0,0000	0,0000	0,0000
Carbohydrate	0,5714	0,5714	0,0113	0,0074	0,0074	0,0000	0,0000	0,0000	0,0000	0,0000
Anion	0,6000	0,4286	0,0113	0,0074	0,0074	0,0000	0,0000	0,0000	0,0000	0,0000
Cation	0,6000	0,4286	0,0113	0,0074	0,0074	0,0000	0,0000	0,0000	0,0000	0,0000
ChemicalElement	0,6000	0,4286	0,0113	0,0074	0,0074	0,0000	0,0000	0,0000	0,0000	0,0000
SemiMetal	0,6000	0,4286	0,0113	0,0074	0,0074	0,0000	0,0000	0,0000	0,0000	0,0000
OrganicMolecule	0,4286	0,4286	0,0113	0,0074	0,0074	0,0000	0,0000	0,0000	0,0000	0,0000
Atom	0,4000	0,2857	0,0113	0,0074	0,0074	0,0000	0,0000	0,0000	0,0000	0,0000
Ion	0,4000	0,2857	0,0113	0,0074	0,0074	0,0000	0,0000	0,0000	0,0000	0,0000
owl:Thing	0,0000	0,0000	0,0113	0,0074	0,0074	0,0000	0,0000	0,0000	0,0000	0,0000

Acknowledgements This work was partially supported by grant RMN-948, research project DOBR/0023/R/ID3/2013/03 and supported by the National Center For Research and Development.

References

1. Chmielewski, M., Gałka, A.: Semantic battlespace data mapping using tactical symbology. Adv. Intell. Inf. Database Syst. **283**, 157–168 (2010). (Studies in Computational Intelligence)
2. Baader, F., Calvanese, D., McGuinness, D.L., Nardi, D., Patel-Sc, P.F. (eds.): The description logic handbook. Cambridge University Press, New York (2010)
3. Chmielewski, M.: Ontology-based indirect association assessment method using graph and logic reasoning techniques. Military University of Technology, Warsaw (2012)
4. Barthelemy, M., Chow, E., Eliassi-Rad, T.: Knowledge representation issues in semantic graphs for relationship detection
5. Stelzl, U., Worm, U., Lalowski, M., Haenig, C., et al.: A human protein-protein interaction network: A resource for annotating the proteome. Cell **122**, 957–968 (2005)

6. Ulrik, B.: A faster algorithm for betweenness centrality. Universität Konstanz, Fachbereich Informatik und Informationswissenschaft. [Online] available http://www.inf.uni-konstanz.de/algo/publications/b-fabc-01.pdf. Accessed 02 June 2012
7. Ehrig, M., Staab, S.: QOM—quick ontology mapping (2004)
8. Kalfoglou, Y., Schorlemmer, M.: Ontology mapping: The state of the art. Knowl. Eng. Rev. **18**, 1–31 (2003)
9. Cormen, T.H., Leiserson, C.E., Rivest, R.L., Stein, C.: Introduction to algorithms, 2nd Edition. MIT Press and McGraw-Hill, pp. 906–932 (2001)
10. OBO—The Open Biological and Biomedical Ontologies webpage. [Online] available http://www.obofoundry.org/. Accessed 15 Nov 2012
11. Stąpor, P.: Ontology structure analysis and visualisation environment. Warsaw, Military University of Technology, Warsaw (2012)
12. Complex Network Parameters - NetworkAnalyzer Online Help. [Online] available http://med.bioinf.mpi-inf.mpg.de/netanalyzer/help/2.5/. Accessed 15 Nov 2012

Decisions as Units of Organizational Knowledge to Manage Complexity

Bart L. MacCarthy and Robert C. Pasley

Abstract Effective decision making can help organizations to manage complexity. Here we argue that considering decisions as units of organizational knowledge and providing a means for decision storage, retrieval and reuse can facilitate effective decision making. To enable the recording and retrieval of decisions, a conceptual model is presented that can be used as a set of requirements for a data structure for decision storage. The approach conforms partly to the proposed Common Decision Exchange Protocol (CDEP) standard, but we extend it to capture decision attributes, decision making stages, decision makers and other collaborators, and the information and tools used. Capturing the linkages between decision elements enables a wide range of organizational decision making processes to be encoded for ontological reasoning. We show the derivation of a new conceptual model from cases. We use the context of decision making in Product Lifecycle Management (PLM) to motivate and illustrate the conceptual model.

Keywords Decisions · Decision making · PLM · Knowledge management

1 Introduction

The literature is weak on viewing decisions as units of organisational knowledge. One schema that has been proposed for storage, reuse and sharing of decisions is the 'Common Decision Exchange Protocol (CDEP)' [1]. This is designed primarily to structure decisions for storage and exchange. CDEP defines a minimum level of structure and fields, but also shows how other aspects such as states of a decision,

B.L. MacCarthy (✉) · R.C. Pasley
University of Nottingham, Nottingham, UK
e-mail: Bart.MacCarthy@nottingham.ac.uk

R.C. Pasley
e-mail: Robert.Pasley@nottingham.ac.uk

© Springer International Publishing Switzerland 2016

J. Świątek et al. (eds.), *Information Systems Architecture and Technology: Proceedings of 36th International Conference on Information Systems Architecture and Technology – ISAT 2015 – Part III*, Advances in Intelligent Systems and Computing 431, DOI 10.1007/978-3-319-28564-1_13

decision makers and supporting information can be stored within the decision. The basic attributes stored about a decision in CDEP [2] are: Question, answer, options (where selection between options occurs), metrics (criteria for the decision), confidence (in the decision), sub-decisions, notes, references (on which the decision is based), states (stages that the decision passes through), basic event information (consisting of "who, what, where, when, how and why"). We argue that CDEP has limitations in the context of organisational and business decision making.

In this paper we propose, explain and justify a minimum sufficient conceptual model for organisational decision making that could be used as a basis for a data structure to support decision capture and reuse. We argue that it provides the basis to improve on the proposed standard. It acknowledges that the organisational context necessarily affects business decisions. We motivate and illustrate these requirements with reference to decision making in the context of Product Lifecycle Managment (PLM) in which multi-disciplinary and hierarchical groups are involved in decision making processes. We demonstrate the relevance of the requirements using hypothetical examples of decision types. We explain how specific requirements for the ontology structure can address issues in these patterns.

2 Product Lifecycle Management

Decisions made during product design and development and later in the Product Life Cycle (PLC) affect how well products are received in the market and ultimately the sustainability of the organisations that design, manufacture and sell them. Many decisions made may not be recorded, and may not be taken in a rational way [2]. Product Lifecycle Management (PLM) refers to organisational processes and systems that support activities across all the phases of the PLC and that enable effective multi-disciplinary collaboration [3]. We derive a high-level conceptual model as a basis for a data structure for use within computer-based PLM systems for the management of the decision making process and also for the recording and storage of decisions for review, modification and reuse across the PLC. The approach has wider applications in the management and support of business and organisational decision making more generally.

Numerous models exist that describe the PLC. For example [4] suggest PLC phases of Definition, Design, Sales, Manufacturing and Service. 'Definition' involves the early stages where the purpose of the product is defined. There may be a high likelihood of termination of a product development project at this stage. At the 'Design' stage the product specification becomes more concrete. This stage can involve team members that help design the product for the various later stages as well as those who further define the product. The 'Sales' stage is where market interest is generated, along with 'Manufacturing', where the product is in production. 'Service' is the continued alteration of the product in order to keep it competitive, and the provision of any support or maintenance needed for the actual product. Towards the end of the PLC the production declines and the product is eventually dropped from the product portfolio. After a product has been discontinued customers may continue to use previously purchased products and such products may require spare parts or servicing. The 'Recycling' phase, where parts and materials may be reclaimed, comes at the end. The naming and granularity of the various PLC phases can vary between PLM models. For example, [5] view the PLC as Beginning of Life (BOL), Middle of Life (MOL) and End of Life (EOL).

The earlier stages of the PLC are described in detail by [3, 6, 7]. Product design has previously been treated as a stage-based process; first the product is defined, and then goes through various stages of design until it has passed through production ramp-up and is available to produce and purchase. This stage-based process is being compressed by companies wishing to reduce time to market [4]. This has resulted in the use of multi-disciplinary teams with much parallelism and interaction between different organizational functions and team members with different backgrounds and expertise [8].

Across the PLC several disciplines can be involved at each phase. For example, design for manufacture seeks to assure that the product is suitable for manufacture and thus requires input from the manufacturing team, and should be suitable for the eventual customer and thus requires input from sales and marketing personnel. Similarly, feedback from existing customers can be used within the design phase. Sales, Manufacturing, Service and Recycling can benefit from knowledge gathered in the earlier stages of the PLC.

PLM is complex from both an organisational and an informational perspective. The complexity cannot be reduced, it can only be managed. PLM complexity is not an artefact of the decision making approach, it is implicit in the context. In order to manage this complexity it is possible to view decisions as necessary elements or branch points within the workflows of normal organisational processes. Decision making needs to be supported for effective PLM.

3 Decision Patterns

Blomqvist et al. [1] use Ontology Design Patterns to develop requirements for CDEP. Below we outline three types of decision context evident from our experience with engineering companies, and outline why the proposed additions to CDEP are necessary and sufficient for similar contexts.

3.1 Broad Consultation Followed by Authorized Decision Making

A recurring pattern is one of consultation with a wide group of advisors and/or stakeholders to gather information, followed by design of a decision process, followed by actual choice between options. For example a product re-design might require input from a number of disciplines. The number and make-up of the stages varies considerably. The early stages may be mainly actioned by people with little decision making authority, but with skills in information gathering or deep experience. The decision design stage may be more suited to mid-level executives who understand decision making, and can select the most salient differentiators between options. At the decision making stage people with authority and responsibility are involved and may be the ultimate arbiters in choosing a course of action. Individuals in the roles mentioned in one decision may well be involved in other decisions in different roles.

3.2 Structuring Decisions

Some decisions are unstructured initially. For example a quality issue arising in an organization may generate considerable debate on its origins and consequences but the decision problem(s) may not be clearly evident. It may be that the decision stages are either undefined or subject to change as the decision proceeds. Supporting information might be used and could be attached to each stage as the decision develops. Information might be emergent from the decision itself, for example the result of brainstorming techniques such as a Delphi study or mind maps might be used to store structures that develop from the decision. These might be useful in later decisions of the same type and as such might be required to be reused. Thus, the decision instance needs to be stored, enabling it to be searched for and retrieved at a later date.

3.3 Multi-criteria Based Decisions

Some decisions are made based on a set of options and a set of criteria. For example a number of different locations might be proposed for a new facility, which may be judged based on a set of criteria such as transport costs, price of land and so on. A typical decision using a multi-criteria decision analysis (MCDA) [9] technique might involve various stages. Firstly, a consultation stage, with potentially many people involved, could explore the issues in such a decision. This would lead to a typical MCDA structure consisting of various criteria and various options and some way to judge these, one against the other. These judgments might be decided by one person or a group of people (who would most likely be different to the group above). In the final stage of the decision someone in authority would approve the decision, most likely following or being informed by the MCDA recommendation. Thus, a schema for representing decisions needs to capture or indicate the information and tools used in different stages of decision making.

4 Proposed Extensions

CDEP has sufficient complexity to deal with decisions taken often where it is clear how the decision is taken and what criteria are used. Typical applications given by [10] are the exchange of decision information during emergencies and also in warfare situations. These reactive situations are in contrast to much organizational and business decision making "where well-defined processes change slowly" [10]. Decision making in an organizational context must consider the organizational hierarchy, organizational protocols and sources of knowledge and information within the organization.

PLM decisions tend to be longer term, often strategic and involve processes that can occur over longer time scales. They require a greater level of structuring than envisaged in CDEP to allow decisions to be taken effectively in a timely fashion, whist adapting to each complex unique situation. The patterns of PLM decisions typically show multiple stages for each decision where these stages may be quite different to each other.

Figure 1 shows the generic context and structure of a decision in the proposed schema. There will always be a context for decision making, either an explicitly defined workflow or an implicit context emerging from experience, expectation or the organization. Decisions fit within this context. In a decision review or decision audit context it is useful to understand how decision points were dealt with over a number of similar instances. For a new decision instance a context can be defined using analysis fields that define the decision attributes. We propose that a generic decision process can be defined by three elements—(1) the people and roles

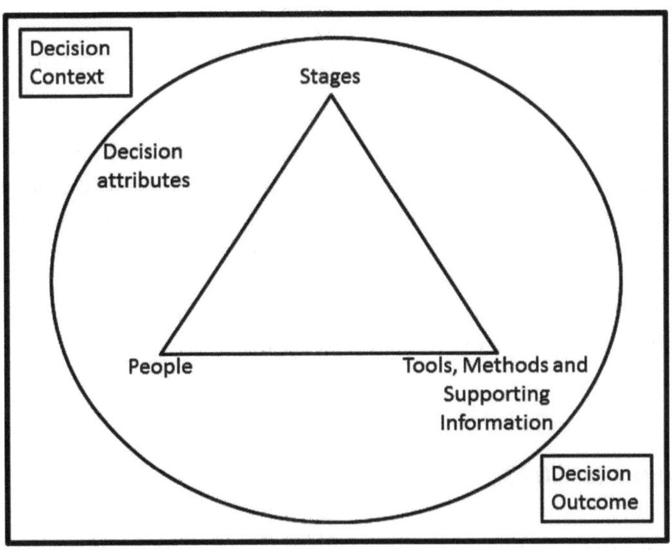

Fig. 1 A conceptual model of the elements of decisions as units of knowledge

involved, (2) the decision stages, and (3) the supporting information and methods, tools and techniques used. We argue that the links between these elements (shown as a triangle) allow the capture of a decision making process in an organizational context. All decisions have an outcome, which also needs to be recorded.

We propose extensions and developments of the CDEP definitions that are necessary to capture the critical decision elements in an organisational decision making context. We identify these as a set of system level requirements for a schema capable of supporting the granularity necessary for decision capture and reuse. This schema supports typical organisational needs: finding suitable people for various stages of decisions; finding decision making process patterns; and finding tools and supporting information for stages of a decision.

4.1 Decision Attributes

We define a set of decision attributes with which a decision can be characterised. These are informed by the decision making literature. Although some will be generic and occur frequently (e.g. urgency of a decision), we propose that they be customizable so that the values used to represent each concept can be changed to suit specific organisations, implementations and contexts. We illustrate these in the context of PLM decision making. See [11] for the derivation of the decision attributes. Here we use the following:

PLM Phase: the point within the PLC at which the decision is being made.
Collaborative: whether the decision is taken by more than one person.
Established or emergent: whether a similar decision has been made before, or if it is the first such decision.
Level and time horizon: whether the decision is strategic, tactical or operational. Level and time horizons are often linked but this attribute may also be disaggregated.
Urgency: an indication of how quickly the decision needs to be taken.
Importance: an indication of the importance of the decision to the organisation.
Structure: whether it is clear how to make the decision based on structured information and methods, or whether the structure needs to be defined.
Project Type: A measure of novelty of the project context in which the decision arises, e.g. PLM project types could be 'Support/enhancement', 'Derivative', 'Platform/next generation' or 'Breakthrough' [12].

4.2 Stages or States

CDEP supports states of decisions, and a set of possible states. We argue that this is a sufficient level of detail to record decision making process workflow but needs to be more formal as many organisations have defined or required stages in making decisions, which may vary by decision type, organisational function or importance. Also, we agree the necessity to capture real timelines (e.g. deadline dates) that can record the actual dates of stage completions. The nature of a decision can be significantly different between stages. For example an 'Information Gathering' stage can be very different to a 'Selection between Options' stage. Different people and different organisational roles may be involved and different types of information may be used to support the decision making process in each stage. Importantly, the decision recording system must also take account of the hierarchical nature of decision making in organisations, as well as the multi-disciplinary nature of decision making evident in many organizations and activities such as PLM.

4.3 People

CDEP supports decision makers, stating that 'A good start is the "who, what, where, when, how and why"' [2]. We take this much further and argue that relating people to decision making roles and to decision making stages enables the capture of a very wide range of organizational decisions and decision processes, the facilitation of such decisions and processes, and further reasoning about decisions. In particular we argue that these linkages are necessary for effective capturing and retrieval of decisions in an organisational or business context.

It is necessary to store the role and who is responsible at each stage of a decision. Such a decision storage system can also give visibility to who the experts are in an organization, which may be useful for instance in building an effective team to address a decision problem. Roles can change over time and between decisions. A specific individual may be a decision maker in one decision but have an advisory role in another. In auditing situations the organization may need to contact people selectively to research decision effectiveness.

4.4 Supporting Information, Tools and Techniques

We expect organisations to take a structuring approach, using different stages of the decision to build on the context of the decision and to clarify key aspects that are important to the decision makers. To this end it is expected that they will use information and tools to support decision stages. In the initial stages such tools, methods and techniques may support ideas about how the decision ought to be structured e.g. supporting brainstorming and mind mapping. As a decision progresses more structure may be evident and tools to support the decision maker(s) in making choices may be appropriate, e.g. MCDA tools, methods and techniques [9]. The information structure proposed should support re-use and reasoning regarding which tools and information sources were used in which stages. There may be enough information about decisions to allow a recommender algorithm to propose or select suitable tools during the decision making process.

To record the content of the tools a snapshot of each tool should be taken. This could be implemented as a simple image in the case of tools like mind maps, or more complex maps for example to show argumentation and reasoning [13]. Tool output relevant to a particular decision stage can be captured when a particular stage finishes, allowing a snapshot to be taken recording the key information.

5 Discussion

The linkages illustrated in the triangle in Fig. 1 provide a powerful and flexible way to capture an instance of organisational decision making. By defining 'stages', 'people/roles', and 'supporting information and tools' and their linkages, we capture how the decision was made. For a specific decision instance, 'Decision attributes' and the 'Outcome' are needed to complete decision capture, defining what the decision was about and result of the decision making process. The stages and the time spent on them, the people involved and tools used, as well as the initial characterisation by decision attributes provide fertile ground for similarity and difference measures to be formulated to support decision retrieval. Decisions may have profiles in terms of these aspects and correlations and associations may become evident between the various aspects.

A likely requirement is to find suitable people and roles for inclusion in a new decision. The retrieval of similar previous decisions may indicate that similar roles or specific people should be involved, including those with specific levels of authority. We argue that in order to satisfy such queries the involvement of each person/role in each decision stage needs to be recorded and stored. When a search is made using the decision knowledge base for responsible authorities to take decisions the results will be focused on the people/roles involved in relevant stages of previous decisions. There is also a potential need to use tools to store supporting information relating to the structure of the decision, such as the options and criteria used. Later there may be a need to use previous experiences with the various different techniques, which can vary in terms of suitability based on the decision context. The structure of each decision is stored by recording the various stages that a decision passes through along with dates and the time that stages change. In this way the structure we propose supports re-use.

6 Conclusions

The proposed extensions to the CDEP standard provide more structure to decision data to enable a wider range of organizational decision contexts to be captured, enabling focused searching and reuse of decisions. We view these as necessary and sufficient for reuse or adaptation of the profile of a previous decision in terms of the decision making process used and in finding appropriate people or supporting information for a new decision. Future work will continue to evaluate the schema in different industrial and business contexts.

References

1. Blomqvist, E., Ceruti, M., Waters, J., McGarry, D.: A decision-making format for the semantic web, p. 53 (2010)
2. Waters, J., Ceruti, M.G., Patel, R., Eitelberg, J.: Decision-acquisition system based on a common decision-exchange protocol (2010)
3. Ameri, F., Dutta, D.: Product lifecycle management: closing the knowledge loops. Comput. Aided Des. Appl. **2**(5), 577–590 (2005)
4. Saaksvuori, A., Immonen, A.: Product lifecycle management. Springer, Berlin (2002)
5. Kiritsis, D.: Semantic technologies for engineering asset life cycle management. Int. J. Prod. Res. **51**(23–24), 7345–7371 (2013)
6. Wheelwright, S.C., Clark, K.B.: Revolutionizing product development: Quantum leaps in speed, efficiency, and quality. Free Press, New York (1992)
7. Kessler, E.H., Chakrabarti, A.K.: Innovation speed: A conceptual model of context, antecedents, and outcomes. Acad. Manage. Rev. **21**(4), 1143+ (1996)
8. Chen, J., Damanpour, F., Reilly, R.R.: Understanding antecedents of new product development speed: A meta-analysis. J. Oper. Manage. **28**, 17–33 (2010)
9. Ishizaka, A., Nemry, P.: Multi-criteria decision analysis: Methods and software. Wiley, Hoboken (2013)
10. Waters, J., Patel, R., Eitelberg, J., Ramstrum, G., Ceruti M.: Information-velocity metric for the flow of information through an organization: Application to decision support. DTIC Document, Tech. rep (2009)
11. MacCarthy, B., Pasley, R.: Decision making and decision support within new product development. In: Collaborative Communication Processes and Decision Making in Organizations. IGI, pp. 258–279 (2014)
12. Hayes, R., Pisano, G., Upton, D., Wheelwright, S.: Operations, strategy, and technology: Pursuing the competitive edge. Wiley, Hoboken (2005)
13. Kunz, W., Rittel, H.: Issues as elements of information systems. Center for Planning and Development Research, University of California at Berkeley (1970)

Decision Making Process for the Knowledge-Based Enterprise: Fuzzy Sets Theory Application to Strategic Management

Joanna Kałkowska and Aleksandr V. Kozlov

Abstract Nowadays, the enterprises are searching for competitive advantage concerning quality of products, application of modern information and telecommunication technologies (ICT), implementation of modern concepts and tools of management as well as proper support of decision making process. To achieve this, enterprises need to possess the ability of the knowledge potential proper usage because widely understood enterprise's development is connected both with the permanent wining, transformation and usage of knowledge and information. This approach allow for the enterprises transformation into the knowledge-based enterprise. The goal of this paper is to determine a fuzzy sets theory to the decision making process for the knowledge-based enterprise management which allows to involve human experience into the process of strategic choice.

Keywords Knowledge-based enterprise · Fuzzy sets · Information and communication technologies (ICT) · Strategic management

1 Introduction

The rapid changes of technology together with the business globalization and proper decision making are considered as a basic factor of an enterprise development and improvement. It also requires the constant knowledge identification and implementation. The organizations structurally subordinate its functionality to

J. Kałkowska (✉) · A.V. Kozlov
Poznan University of Technology, Poznań, Poland
e-mail: joanna.kalkowska@put.poznan.pl

A.V. Kozlov
Peter the Great Saint-Petersburg Polytechnic University, Saint Petersburg, Russia
e-mail: avk55-spb@yandex.ru

© Springer International Publishing Switzerland 2016

J. Świątek et al. (eds.), *Information Systems Architecture and Technology: Proceedings of 36th International Conference on Information Systems Architecture and Technology – ISAT 2015 – Part III*, Advances in Intelligent Systems and Computing 431, DOI 10.1007/978-3-319-28564-1_14

create the added value on the basis and with the use of knowledge resources. The knowledge is the key resource of the organization. The efficient usage and knowledge implementation is possible through the information technologies application that allows both the knowledge codification and dissemination. Specifics of knowledge-based enterprises define needs in the new management approaches to be used by ICT. Approach to decision making based on fuzzy sets theory allows to involve human experience into decision making process and intellectualize it. Capability of this approach to work is to be proved by application of it to strategic management. The goal of this paper is to demonstrate a fuzzy sets theory approach to decision making process for knowledge-based enterprise management allowing to involve human experience into the process of strategic choice.

2 Modern Economy and Knowledge Based Organization—Fundamentals

The widely understood knowledge became a fundamental driver of increased productivity and global competition. Knowledge is also the most valuable source of competitive advantage as well as it is considered to be the prominent resource of enterprise in terms of its contribution to the value added and its strategic significance [2]. However, the dominance of the knowledge in the social and economic life resulted in the nineties of the 20th century with introducing into the economic theory and practice is the concept of "knowledge based economy" or so called "modern economy". Ambiguity in understanding the knowledge based economy causes that many sets of features with different degree of accuracy are being used in descriptions of this phenomenon. The classic definition presented in 1996 by the OECD shows a knowledge-based economy as an economy which directly is based on the production, distribution and using the knowledge and the information [16]. According to the World Bank, the idea of knowledge based economy claiming that the economy is becoming an "knowledge based economy", when using and creating knowledge maintain permanently the center of its processes of economic development. A knowledge-based economy is an economy which uses knowledge as a motor of economic growth [19]. Knowledge is treated here as a fundamental driving force of the economy, as a factor stimulating to progress. The knowledge economy development is mainly connected with the two premises. First of it is connected with the growing importance of the knowledge as a dominant resource of social and economic development—the development of the information society, the growing importance of intellectual capital and the development of information and communication technologies (ICT) an their increase in the efficiency of their use in society favoring processing and management of knowledge resources. The second one is focused on the intensive growth and dominance in the economy processes causing an increase of the knowledge importance while undertaking the management decisions. Among them the most important are the following: focus on the customer, the globalization of markets, turbulence, the changeability and uncertainty of the environment which requires the changes, flexibility and innovations as well as the intensive usage and development of information and communication technologies (ICT) in management processes and while undertaking the management decisions.

According to this, it have to be considered that one of the most important factor of socio-economy development as well as improvement of competitiveness became an enterprise's transformation into requirements of knowledge based economy. The enterprises are searching for competitive advantage concerning quality of products, manufacturing costs, time of launching products, application of modern information technologies and using actual and valuable information. To achieve this, first of all enterprises need to possess the ability of the knowledge potential proper usage because widely understood enterprise's development is connected both with the permanent wining, transformation and usage of knowledge and information [7].

The development of knowledge-based organization is observed from the beginning of 90 years. Characteristics of this concept development were described repeatedly in literature [4, 5, 8, 13, 14].

Summarizing the characteristics of the knowledge-based organizations it is possible to recognize the following features (Fig. 1) [14]:

1. The structure of resources and investments in immaterial sources which constitute the majority element of the organization's market value. In particular it is about an intellectual capital which includes:

   - Human capital (people and their knowledge, abilities, values, norms, attitudes, opinions, emotional intelligence, etc.),
   - Structural capital—understood mainly as an organizational capital, created by the processes, internal and external, used methods, software, databases and documents;
   - Customer capital created by customers. It reflects their potential value of the purchased products and services offered by the organization,
   - Intellectual property, including patents, licenses, copyright, trademarks, projects, etc.

2. Knowledge management understood as conscious and intentional knowledge management, including aspects of strategy, structure, culture, technology and people.
3. Shaping the relation with the environment using one's knowledge to get the beneficial localization in the economy network. These relations are also a base of the knowledge transfer from the organization's environment, confronting with own knowledge and creating new knowledge resources.
4. Organizational structure characterized by the high flexibility, openness for environment in frames of network and virtual structures, the broader exploitation of temporary teams and forming positions or teams for the knowledge management.

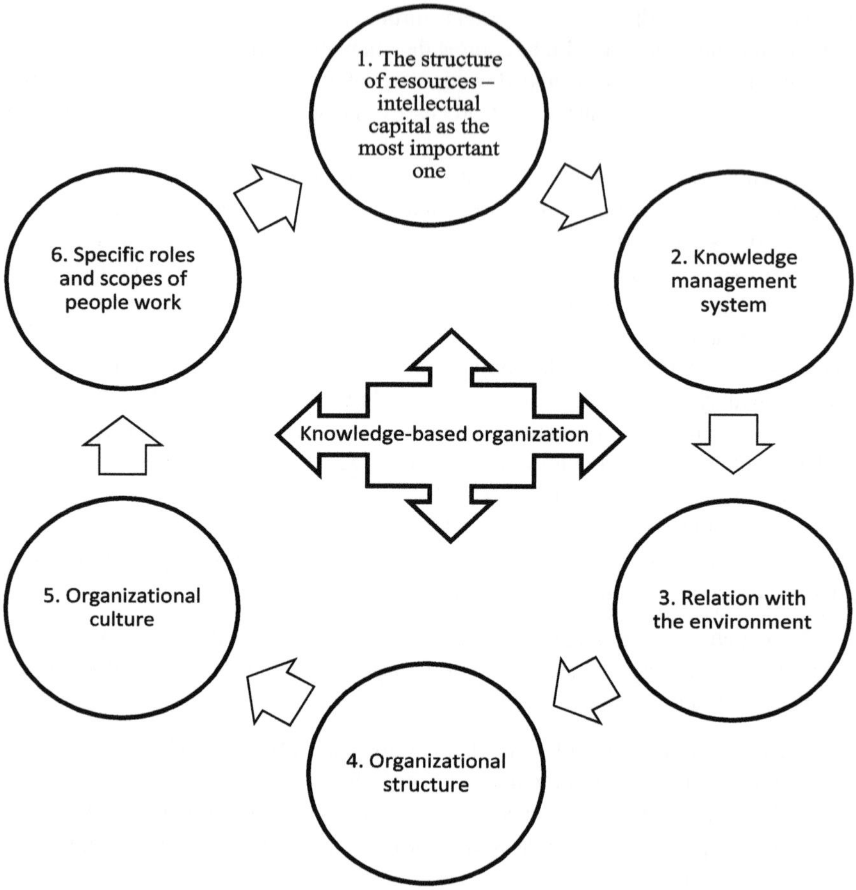

Fig. 1 The main features of the knowledge-based organization (*source* own study on the basis on: 7)

5. Organizational culture adapted for the new conditions and supporting the knowledge management: intellectual programs of people behaving supporting knowledge creation and transfer, learning culture, team working as well as the mutual confidence in human relationships.
6. Specific roles and scopes of people work. Widening the range and scope of people activities as well as the pressure on the initiative and searching for the system improving possibilities.

The set of above features of the knowledge-based organization has evolved for years and still doesn't constitute the uniform concept, although it seems that divergences aren't widening indeed in a few last years. These features are certain aggregate definitions with the unclear border. For example, the sixth feature: specific roles and scopes of people work seem to be a part of the fourth feature:

organizational structure. It is also obvious, that it is not only point of view, and intellectualizing of variables being characteristic of an examined object depends on a view point and the purpose of research. The existing diversity of the concept is a natural consequence of the system approach to the scientific researches. The whole of descriptions of the object from the different viewpoints gives the quite comprehensive image. The described above set of six categories of the knowledge-based organization has the model character containing features universally recognized of such an organization [8].

3 Information and Communication Technologies (ICT) and Decision Making Process

Information technologies are a critical factor for the effective operation and prosperity of modern organizations. However, management and dissemination of information is a central for the enterprise [15]. Also other result of ICT application in enterprise is depart from hierarchical organizational structure which very often lead to communication barriers, mistakes in information flow as well as wasting human knowledge [11]. Moreover, ICT are considered as a one of the most important factors of modern enterprises development and competitiveness. ICT usage influence also on a number of indicators allowing for competitiveness estimation. These indicators are following: costs reduction, quality improvement, increasing production flexibility, products and technology innovativeness, extending products assortment and productivity, introducing into international markets [3]. Moreover, thanks to the information infrastructure and proper software it is possible a quite fast coordination of all tasks carried out in enterprise. Besides, one of key factors of enterprise's management system in a context of knowledge based economy is including information technologies into widely understood decision making and management processes. In that meaning, these technologies are treated as one of determinants of enterprises' transformation into knowledge based economy, enabling not only for increasing but also for creating of new organizational knowledge. Furthermore, information technologies may dramatically enhance the coordination and control capacity of the enterprise, so that way can stimulate increased use of management system. ICT removes distance and time constrains in accessing required information flows and hence improves the coordination of activities within organizational boundaries. ICT affects planning systems by improving organizational communication as well as by enhancing organizational flexibility [18], however, the application of advanced IT corresponds to an incremental process of organizational capability and strategic impact [18].

If we generalize well-known attitudes to the structure of decision making process we may identify following steps: (1) Identification of the problem and goal setting, (2) Information gathering, (3) Alternative decisions development, (4) Choice of the decision based on the criteria formulated, (5) Decision implementation and control, (6) Results valuation.

ICT have an impact on process of making of managerial decisions on every step proposing standard or adopted procedures and techniques. Nevertheless we can't apply standard ICT decision making procedures to management of knowledge-based enterprises because of above mentioned peculiarities. Actually every step needs to be changed since it's necessary to use decision making tools adequate to the object to be managed. It's necessary to intellectualize these ICT procedures by involving into decision making an experience of managers. There are few tools allowing to incarnate human experience into ICT procedures. Such a tool which got a title of fuzzy sets theory was proposed by Zadeh [21].

4 Fuzzy Sets as a Tool for Decision Making: Example of Strategic Management

The process of strategic management decision making may be presented as a graph (Fig. 2). Every vertex means a point of strategic decision making where choice among some number of alternative strategies is available. If strategy Sij will be chosen at vertex "i", it will be implementing during definite period of time up to the moment when monitoring of business environment will show a necessity of strategy change or modification (vertex j, Fig. 2). A new choice is to be done by enterprise.

Let's propose that there are three possible scenarios of the external environment future development in vertex "i" of the graph: optimistic, pessimistic and moderate. Top management of the enterprise has developed three strategies to choose for every scenario. Let's assume that probability of optimistic, moderate and pessimistic scenarios will be $\boldsymbol{p_i^o}$, $\boldsymbol{p_i^m}$ and $\boldsymbol{p_i^p}$, correspondingly (Fig. 3).

The task will be formulated as follows: to determine $\boldsymbol{p_i^o}$, $\boldsymbol{p_i^m}$ и $\boldsymbol{p_i^p}$ dependently on the results of analysis of external business environment done by experts.

$$p_i^o + p_i^m + p_i^p = 1 \tag{1}$$

$$p_i^o \geq 0;\ p_i^m \geq 0;\ p_i^p \geq 0, \quad i \in 1, n;$$

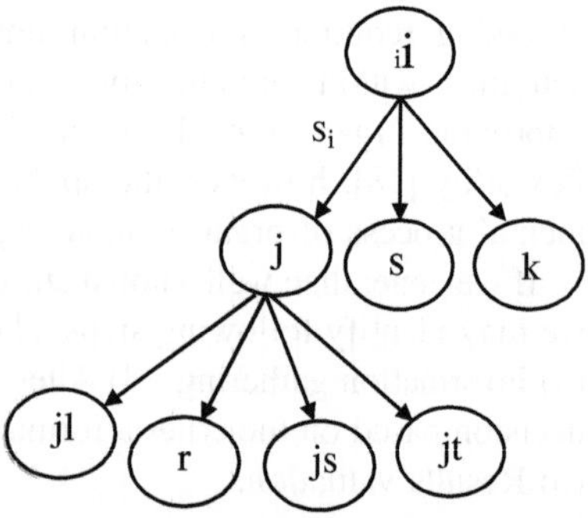

Fig. 2 The process of strategic management decision making (*source* own study)

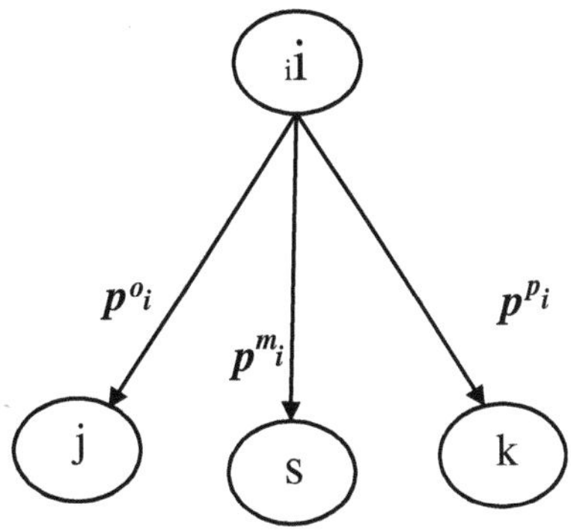

Fig. 3 Probability of optimistic, moderate and pessimistic scenarios (*source* own study)

Intellectualizing of decision making process means into this process of human experience through involvement of qualified experts [17]. The most convenient way for experts to assess opportunities and threats presented by an environment is to use linguistic scale as proposed below:

1. Opportunities are dominating, no principal threats
2. More opportunities fewer threats
3. Opportunities and threats are equal
4. More threats fewer opportunities
5. Threats are dominating, no principal opportunities

There are an essential number of tools for external analysis, check-lists and questionnaires, based on different methodologies: SWOT-analysis, PEST-analysis [9], PESTEL-analysis, GETS model, etc. Top managers of an enterprise as experts may use any proper tool for correct and comprehensive assessment. Thus the main task is to transform qualitative results of assessment located on linguistic scale into quantitative figures, to digitalize it. The idea of digitalization rests on fuzzy sets theory.

The idea of fuzzy sets theory was proposed by Zadeh [21]. A lot of different applications of fuzzy sets were developed which proved reliability of this tool. Generally speaking a real fuzzy number N [α, β, γ] is an interval around the real number β with the elements in the interval being partially present. $\mu_N(x)$ is $\Psi_1(x)$, if $\alpha \leq x \leq \beta$, is $\Psi_2(x)$, if $\beta \leq x \leq \gamma$, and is 0, otherwise [1]. Partial presence of an element is defined by membership function μ_N (β): $\rightarrow$ [0, 1], where [0, 1] means $0 \leq \beta \leq 1$. The other way to describe fuzzy set N including membership function is the following: $N = \{\beta, \mu_N(\beta)\}$.

A lot of different applications of fuzzy sets theory were proposed for different managerial [20], economic [6], [12] and engineering [10] tasks. Nevertheless there are few researches devoted to application of fuzzy sets to strategic management.

Table 1 Recommended fuzzy scale for linguistic variable Z

#	Expert assessment	Scale
1	Threats are dominating, no principal opportunities	$0 < z < 0.333$
2	More threats fewer opportunities	$0.167 \le z < 0.5$
3	Opportunities and threats are equal	$0.333 \le z < 0.667$
4	More opportunities fewer threats	$0.5 \le z < 0.833$
5	Opportunities are dominating, no principal threats	$0.667 \le z \le 1$

Source own study

Let introduce linguistic variable Z describing the relationship between opportunities and threats as it's presented in Table 1.

Let introduce d_j—result of assessment of an expert j ($d_j \in 1, 2, 3, 4, 5$). According to the fuzzy set theory membership for set N can be expressed through $\mu_N(x)$—membership function with an interval [0, 1]. It can be described for every d_j by the following fuzzy sets N_s, s = 1, 2, 3, 4, 5, according to expert's choice.

$$N_1 = \{(\beta_1, \mu_N(\beta_1)), (\beta_2, \mu_N(\beta_2)), (\beta_3, \mu_N(\beta_3))\}, \text{ if } d_j = 1; \tag{2}$$

$$N_2 = \{(\beta_2, \mu_N(\beta_2)), (\beta_3, \mu_N(\beta_3)), (\beta_4, \mu_N(\beta_4))\}, \text{ if } d_j = 2; \tag{3}$$

$$N_3 = \{(\beta_3, \mu_N(\beta_3)), (\beta_4, \mu_N(\beta_4)), (\beta_5, \mu_N(\beta_5))\}, \text{ if } d_j = 3; \tag{4}$$

$$N_4 = \{(\beta_4, \mu_N(\beta_4)), (\beta_5, \mu_N(\beta_5)), (\beta_6, \mu_N(\beta_6))\}, \text{ if } d_j = 4; \tag{5}$$

$$N_5 = \{(\beta_5, \mu_N(\beta_5)), (\beta_6, \mu_N(\beta_6)), (\beta_7, \mu_N(\beta_7))\}, \text{ if } d_j = 5, \tag{6}$$

for j = 1, 2, 3,.., E,

E—number of experts assessing the environment.

Thus quantitative description of the fuzzy sets Ns including membership functions for one expert are presented at Table 2.

Let's take the following distribution of domains and corresponding scenarios of the future development of business environment:

Table 2 Membership functions for fuzzy sets

Fuzzy set	Element of fuzzy set and membership function for fuzzy set element						
N_1	0, 0.5	0.167, 1.0	0.333, 0.5	0.5, 0.0	0.667, 0.0	0.833, 0.0	1.0, 0.0
N_2	0, 0.0	0.167, 0.5	0.333, 1.0	0.5, 0.5	0.667, 0.0	0.833, 0.0	1.0, 0.0
N_3	0, 0.0	0.167, 0.0	0.333, 0.5	0.5, 1.0	0.667, 0.5	0.833, 0.0	1.0, 0.0
N_4	0, 0.0	0.167, 0.0	0.333, 0.0	0.5, 0.5	0.667, 1.0	0.833, 0.5	1.0, 0.0
N_5	0, 0.0	0.167, 0.0	0.333, 0.0	0.5, 0.0	0.667, 0.5	0.833, 1.0	1.0, 0.5

Source own study

Table 3 Pessimistic scenario zone inside of matrix of membership functions for fuzzy sets

Fuzzy set	Element of fuzzy set and membership function for fuzzy set element					
N_1	*1, 0.5*	*0.167, 1.0*	*0.333, 0.5*	0.5, 0.0	0.333, 0.0	0.167, 0.0	1.0, 0.0
N_2	*0, 0.0*	*0.167, 0.5*	*0.333, 1.0*	0.5, 0.5	0.333, 0.0	0.167, 0.0	1.0, 0.0
N_3	*0, 0.0*	*0.167, 0.0*	*0.333, 0.5*	0.5, 1.0	0.333, 0.5	0.167, 0.0	1.0, 0.0
N_4	*0, 0.0*	*0.167, 0.0*	*0.333, 0.0*	0.5, 0.5	0.333, 1.0	0.167, 0.5	1.0, 0.0
N_5	*0, 0.0*	*0.167, 0.0*	*0.333, 0.0*	0.5, 0.0	0.333, 0.5	0.167 1.0	1.0, 0.5

Source own study

Table 4 Moderate scenario zone inside of matrix of membership functions for fuzzy

Fuzzy set	Element of fuzzy set and membership function for fuzzy set element						
N_1	1, 0.5	0.167, 1.0	*0.333, 0.5*	*0.5, 0.0*	*0.333, 0.0*	0.167, 0.0	1.0, 0.0
N_2	0, 0.0	0.167, 0.5	*0.333, 1.0*	*0.5, 0.5*	*0.333, 0.0*	0.167, 0.0	1.0, 0.0
N_3	0, 0.0	0.167, 0.0	*0.333, 0.5*	*0.5, 1.0*	*0.333, 0.5*	0.167, 0.0	1.0, 0.0
N_4	0, 0.0	0.167, 0.0	*0.333, 0.0*	*0.5, 0.5*	*0.333, 1.0*	0.167, 0.5	1.0, 0.0
N_5	0, 0.0	0.167, 0.0	*0.333, 0.0*	*0.5, 0.0*	*0.333, 0.5*	0.167 1.0	1.0, 0.5

Source own study

(0, 0.333)—pessimistic scenario (see Table 3—a part a presented as an italic font);
(0.333, 0.667)—moderate scenario (see Table 4—a part a presented as an italic font);
(0.667, 1.0)—optimistic scenario (see Table 5—a part a presented as an italic font).

In case if we take into account that we use normalized fuzzy set, we receive the following scenario zone inside of matrix of membership functions for fuzzy sets.

Let's introduce r_{sf} as an element of a fuzzy set s and q_{sf} as a membership function for corresponding fuzzy set element at matrix presented at Table 2, we may count probability for one expert assessment (generalized assessment of few experts) in a following way.

Step 1. Counting of Pi—intermediate assessment of probability, i = 1, 2, 3., reflecting a frequency of choice done by expert:

$$P_1 = \sum_{f=1}^{3} r_{1f} q_{1f};\ P_2 = \sum_{f=3}^{5} r_{2f} q_{2f};\ P_3 = \sum_{f=5}^{7} r_{3f} q_{3f}. \quad (7)$$

Table 5 Optimistic scenario zone inside of matrix of membership functions for fuzzy

Fuzzy set	Element of fuzzy set and membership function for fuzzy set element						
N_1	1, 0.5	0.167, 1.0	0.333, 0.5	0.5, 0.0	*0.333, 0.0*	*0.167, 0.0*	*1.0, 0.0*
N_2	0, 0.0	0.167, 0.5	0.333, 1.0	0.5, 0.5	*0.333, 0.0*	*0.167, 0.0*	*1.0, 0.0*
N_3	0, 0.0	0.167, 0.0	0.333, 0.5	0.5, 1.0	*0.333, 0.5*	*0.167, 0.0*	*1.0, 0.0*
N_4	0, 0.0	0.167, 0.0	0.333, 0.0	0.5, 0.5	*0.333, 1.0*	*0.167, 0.5*	*1.0, 0.0*
N_5	0, 0.0	0.167, 0.0	0.333, 0.0	0.5, 0.0	*0.333, 0.5*	*0.167 1.0*	*1.0, 0.5*

Source own study

Step 2. Counting of pi—probabilities of pessimistic, moderate and optimistic scenarios, i = 1, 2, 3, correspondingly.

To count probabilities we need no normalize the intermediate assessment of probabilities according to following formula

$$\mathrm{p}_i = P_i \Big/ \sum\nolimits_{\mathrm{i}=1}^{3} P_i. \tag{8}$$

Check capacity of the model to work by counting values of pi for practical case of choice of expert equal 3. In that case counting will be as follows

$$\begin{aligned} P_1 &= (0 \times 0.0 + 0.167 \times 0.0 + 0.333 \times 0.5) = 0.167; \\ P_2 &= (0.333 \times 0.5 + 0.5 \times 1.0 + 0.333 \times 0.5) = 0.833; \\ P_3 &= (0.333 \times 0.5 + 0.167 \times 0.0 + 1.0 \times 0.0) = 0.167. \end{aligned} \tag{9}$$

The normalization gives following values of probabilities

$$\mathrm{P}_1 = 0.143;\ \mathrm{p}_2 = 0.714;\ \mathrm{p}_3 = 0.143. \tag{10}$$

Thus, the most probable scenario in given case is the moderate one. That fact gives to top managers of an enterprise crucial about the future.

Limitations:
As stated in [1] all of fuzzy sets theory applications are to be proved by practice. This particular application isn't an exception. Only comparison of results of modelling with practice can prove how good this model fits to real practice of strategic decision making. Moreover we should hope that management of enterprise is qualified enough to be able to develop alternative strategies for different scenarios of future business environment development. One more limitation is connected with number of experts involved into assessment of business environment and number of factors taken into consideration by the experts. A simplified model of strategic choice is described in the paper based on model of only expert involved (or few experts proposing agreed opinion). Number of factors is also limited by the only one (or reduced from few to agreed one by experts).

5 Conclusions

Knowledge-based enterprise management needs an adequate decision making system which fits to modern challenges of global economy. The most convenient system for knowledge based enterprise intellectual one. The process of intellectualization in strategic decision making is connected with involvement of human experience. The paper presents a tool for intellectualization of strategic decision making process based on fuzzy sets theory. The fuzzy sets approach allowed implementing human expertise into computer based information technologies. Practical sample of counting of value of probabilities for three alternative scenarios of the future business environment development confirmed a capacity of the model to work.

References

1. Baruah, H.K.: The theory of fuzzy sets: Beliefs and realities. Int. J. Energy Inf. Commun. **2**(2), 1–22 (2011)
2. Bolisani, E., Scarso, E.: Information technology management: A knowledge-based perspective, Technovation no. 19, Elsevier Science Ltd., Amsterdam (1999)
3. Chlebus, E.: Zintegrowany rozwój produktu i procesów produkcyjnych. Oficyna Wydawnicza Politechniki Wrocławskiej, Wrocław (2002)
4. Grudzewski, W.M., Hejduk, I.K.: Zarządzanie wiedzą w przedsiębiorstwach. Difin, Warszawa (2004)
5. Jashapara, A.: Zarządzanie wiedzą. PWE, Warszawa (2006)
6. Jiang, G.-G.: Empirical analysis of regional circular economy development-Study based on Jiangsu, Heilongjiang, Qinghai Province (Conference Paper). 2010 International Conference on Energy, Environment and Development, ICEED 2010, vol. 5, pp. 125–129. Energy Procedia, Kuala Lumpur (2011)
7. Kałkowska, J., Włodarkiewicz-Klimek, H.: Information technologies in concurrent engineering in selected knowledge based economy companies in Poland—some results of research. In: Karwowski, W., Salvendy, G. (eds.) Advances in Human Factors, Ergonomics and Safety in Manufacturing and Service Industries, pp. 97–107. Taylor and Francis Group, Boca Raton (2011)
8. Kałkowska, J., Pawłowski, E., Włodarkiewicz-Klimek, H.: Zarządzanie organizacjami w gospodarce opartej na wiedzy. Wydawnictwo Politechniki Poznańskiej, Poznań, 131 s (2013)
9. Kałkowska, J., Pawłowski, E., Trzcielińska, J., Trzcieliński, S., Włodarkiewicz-Klimek, H.: Zarządzanie strategiczne: metody analizy strategicznej z przykładami. Wydawnictwo Politechniki Poznańskiej, Poznań (2010)
10. Klir, G.J., Yuan, B.: Fuzzy sets and fuzzy logic. Theory and applications, p. 574. Prentice Hall, New York (1995)
11. Kobyłko, G., Morawski, M. (eds.): Przedsiębiorstwo zorientowane na wiedzę. Wydawnictwo Difin, Warszawa (2006)
12. Kozlov, A., Gutman, S., Rytova, E., Zaychenko, I.: The application of the fuzzy set theory to counting a regional innovative development indicators: The case of the Yamal region of Russian Federation. In: Proceedings of IEEE XVIII conference in Soft Computing and Measurements. SCM'2015, St. Peteresburg (2015)
13. Mikuła, B.: Organizacje oparte na wiedzy. Wydawnictwo Akademii Ekonomicznej, Kraków (2006)
14. Mikuła, B., Pietruszka-Otyl, Potocki, A.: Podstawy zarządzania organizacjami w gospodarce opartej na wiedzy. Wydawnictwo Difin, Warszawa (2007)
15. Morgan, A., Colebourne, D., Brychan, T.: The development of ICT advisors for SME business: an innovative approach, Technovation no 26, pp. 980–987 (2006)
16. OECD: The knowledge-based economy. OECD, Paris (1996)
17. Reyna, V.F.: A new intuitionism: Meaning, memory, and development in fuzzy-trace theory. Judgement Decis. Mak. **7**, 332–359 (2012)

18. Spanos, Y.E., Prastacos, G.P., Poulymenakou, A.: The relationship between information and communications technologies adoption and management. Inf. Manag. **39**, 659–675 (2002)
19. Wei, G., Peng, J.: An approach of fuzzy comprehensive evaluation on supporting capacity of regional science and technology with case study of Chongqing (Conference Paper). In: International Conference on Management and Service Science. MASS, pp. 1–5 (2011)
20. World Bank: Korea as knowledge economy. Evolutionary process and lessons learned. Overview. World Bank, Washington (2006)
21. Zadeh, L.: Fuzzy sets. Information and control, pp. 338–353 (1965)

Part IV
Knowledge Based Management

Computer Science PhD Program Evaluation Proposal Based on Domain and Non-domain Characteristics

Jan Werewka and Michał Turek

Abstract There is increasing interest in PhD studies, also referred to as 3rd level studies. PhD studies for employees in the IT sector should be mostly focused on strategic fields, advances in software engineering, and software system engineering, however IT architecture should also be an important part of these programs. This evaluation of PhD programs is concerned with their width, depth, quality, and value to the IT industry. It is important to distinguish domain and non-domain characteristics when designing PhD studies. This paper is concerned with establishing a part-time doctoral program in computer science in the context of cooperation with the IT industry. The results of analysis and evaluation of part-time studies are described.

Keywords Phd studies · IT architecture · Software architecture · Computer science · Education

1 Introduction

In the IT sector there is a demand for experts who are able to take a broad view and recognize the whole spectrum of issues, while at the same time being capable of conducting in-depth scientific research. The proposed part-time PhD program allows its participants to reconcile professional activity with developing valuable new skills.

J. Werewka (✉) · M. Turek
Department of Applied Computer Science, AGH University of Science and Technology, al. Mickiewicza 30, 30-059 Kraków, Poland
e-mail: Jan.Werewka@agh.edu.pl

M. Turek
e-mail: mitu@agh.edu.pl

© Springer International Publishing Switzerland 2016

J. Świątek et al. (eds.), *Information Systems Architecture and Technology: Proceedings of 36th International Conference on Information Systems Architecture and Technology – ISAT 2015 – Part III*, Advances in Intelligent Systems and Computing 431, DOI 10.1007/978-3-319-28564-1_15

A PhD program in IT engineering provides an opportunity to develop engineers with the advanced knowledge and innovative skills which benefit the IT sector. In the "Bologna Declaration" [1] the European Union (EU) promotes consistency and mobility in the field of higher education. The Declaration introduced a framework for achieving compatibility by recommending a clear demarcation between undergraduate degree studies (1st level, at least 3 years of coursework), graduate studies (2nd level, 2 years of coursework), and doctoral degrees (3rd level, 3 years). Report [2] ascertains that the US system of graduate education is a strategic national asset. This report examines the data behind this assertion, and proposes a set of recommendations to strengthen U.S. graduate (including PhD) education in partnership with industry and government.

The course is organized by AGH University of Science and Technology, Faculty of Electrical Engineering, Automatics, Computer Science and Biomedical Engineering, Department of Applied Computer Science. The three-year, part-time IT PhD program is based, on average, on 16 weekend sessions per year in the first two academic years, beginning in October and ending in June. The third year concentrates on PhD thesis consultancy and passing doctoral exams.

The goal of the project presented in this paper is to propose PhD studies for professionals specializing in IT architecture who work in the IT sector and the computer science domain. The term IT architecture should be understood broadly and means developing architecture, including enterprise, business, application, and infrastructure architectures. For the sake of clarity, we are considering establishing part-time PhD studies in the field of computer science, specializing in IT Architecture. The studies have the following distinctive features:

- Weekend courses allow participants to reconcile the development of knowledge about information technologies with their professional work.
- Addressed to practitioners who are currently building their professional career outside university and are preparing for its dynamic future growth.
- In the course of studies, students will be provided with the opportunity to consolidate their horizontal knowledge about creating efficient IT systems
- Intensive teaching program for the first two years. The first two years of the course are intended to build expert knowledge about ICT Architecture and provide the foundations for conducting scientific work.
- Even those students who decide to leave the course after the first or the second year will gain specialist knowledge of great value in the IT sector.
- The knowledge gained by students will be coherent and will significantly increase the their value on the employment market.
- The international character of the studies promotes the sharing of experiences and cooperation which are required globally.
- The implementation of e-learning will facilitate team work.
- The studies encourage assessment of the business value of developed solutions and projects.

- The studies support scientific work which addresses the challenges related to the development of modern technologies and the necessity to develop various methodologies and optimization of solutions at the appropriate level of abstraction.

Assuming there is a need for such studies, the following issues arise: (1) how should the studies be organized, so as to be suitable for employees in the IT sector? (2) How to identify the PhD studies' areas of education which are particularly relevant for cooperation with industry (3) How to obtain high quality studies (4) How should cooperation with industry be structured to give significant benefits to both university and industry? These questions are answered in this paper by investigating domain and non-domain features of the studies.

2 Related Works

The structure of PhD studies was investigated from different perspectives. In paper [3] a study was performed to understand and develop a theory explaining the process domestic engineers undergo when developing an interest in obtaining a PhD in engineering. As a result of the study a framework was provided for understanding and promoting doctoral education for engineers. Several universities have established or are considering establishing engineering education research centers and PhD programs [4] and have set up a space on the Collaboratory for Engineering Education Research (CLEERhub.org) for information exchange. The ELLEIEC project, which is focused on Electrical and Information Engineering, analyzed the situation of PhD students in Europe with the goal of highlighting the differences between the delivery of diplomas and the adequacy of academic institution regulations [5]. The same project is intended [6] to highlight several aspects of the current situation and proposes recommendations in order to facilitate mobility and exchange in Europe at PhD level. The final part of this work is devoted to recommendations to European partners in order to improve the present situation and create a real European space for doctoral studies. Paper [7] presents an innovative project financed by the European Union that aims to promote quality research in aeronautics and global PhD education by fostering and financing Doctoral (PhD) level research, involving the participation of leading universities, research centers and the Air Traffic Management industry, through a collaborative research network. The information economy requires skilled industry-based professionals who can communicate effectively and contribute to the global economy [8]. Moreover, many industries require innovative strategic leaders with solid theoretical and research grounding. Such leaders draw on their education and experience to develop strategically aligned technical communication practices for specific domains and industries. The research revealed potential gaps in the availability of PhD educational opportunities for industry professionals who must work full-time whilst still in education. In IT companies, IT architects play a strategic role in the development of software and in offering a valuable product range. Study [9] shows a systematic approach to the development and assessment of competences which could be leveraged in software development companies.

3 PhD Studies Domain Versus Non-domain Characteristics

General studies are characterized by domain and non-domain features. There are 4 main reference points of PhD study evaluation: pre-conditions, area of studies, research agenda, and post-conditions (Fig. 1).

The preconditions for applicants are as follows. The applicants should be graduates of second-cycle university programs in the field of technical sciences, economics, or natural science, and show the ability necessary to continue education at the level of doctoral studies. They should have an interest in IT research, a systematic approach, the ability to formulate research problems, and practical experience in software development or the application of information technology. Applicants who fall well short of the preconditions should attend compensatory classes to gain ECTS level 6.

This may be achieved by attending classes at the IT Architecture Academy [10], which offers education paths following SEI, ISAQB and Open Group approaches.

Post-conditions state that a PhD degree in computer science is a confirmation of scientific achievements in computer science and expert knowledge in the domain of the PhD studies.

The other reference points of the study evaluation are described in subsequent sections.

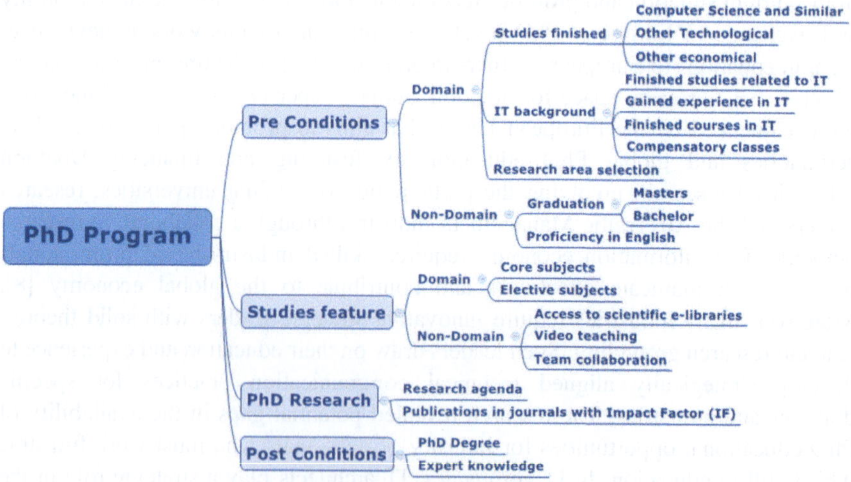

Fig. 1 PhD Studies domain versus non-domain characteristics

4 Non-domain Characteristics Evaluation of PhD Studies

PhD studies should be well-organized, intensive and time efficient. A vast amount of research needs to be done in order to define the set of features important for evaluating studies. There are many sources of knowledge regarding the subject. A good and reliable way to find data is to process multiple search results using phrases from a kind of dictionary of the most common features. A similar technique was used for determining non-domain attributes for certified courses [11].

Evaluation basically assumes the establishment of a set of preliminary studies together with a huge set of possibly important study features. Subsequently, an iterative feature set improvement process starts. To achieve this, a special dictionary is established, containing synonyms of features, prefixes, negation phrases, and so on. A modified standard Google search robot was used to conduct an automated cross-tabulate search procedure.

Subsequently, query result counts were collected for words based on features with the use of a preliminary set of PhD studies names.

Table 1 Query results of hit-counts for non-domain features

	Non-domain feature	Counts
1	Graduation requirement for Admission Bachelor or Master's degree	37,132
2	IT or similar graduation requirement for admission	10,032
3	Admission exam	4021
4	Proficiency in English needed for admission	503,123
5	Face-to-face presence required	98,110
6	Online (e-learning) courses	930,124
7	Foreign students allowed	600,416
8	Part-time studies mode allowed	9020
9	Obligatory number of hours	80,213
10	Presence of education points (ECTS)	77,610
11	Studies specialization options (Tracks)	309,102
12	Tuition fee	60,430
13	Access to on-line libraries	900,812
14	Fixed length of the studies	60,133
15	PhD thesis fee	90,902
16	Dissertation proposal as a conditions to open PhD thesis	4012
17	Presence of dissertation committee	1273
18	PhD thesis is a condition to defend	10,234
19	PhD thesis public (advertised) defense	40,164
20	Publications as progress measures	86,900
21	Careers perspective defined	66,012
22	Partnership with industry for research on real projects	4051
23	Presence of interdisciplinary levels (coming beyond IT)	2201
24	Predefined studies program (same for all students)	5301

Table 1 shows sample results listed with the hit count for each non-program feature. The total on the left explicitly represents positive search results for each query.

This is simply a best-effort method for evaluating non-domain features, however constant improvement means the process must be started with updated data each time someone wants to evaluate a new PhD program.

An in-depth analysis of the PhD studies market and numerous case studies based on individual certification schemes helped when choosing non-program features and rank in terms of the number of appearances. The results obtained from search engines will be the basis for initial surveys with employees in the IT sector.

5 Domain Characteristics Evaluation of PhD Studies

Areas of the Part-time Doctoral Studies Program are related to solutions which are of current value for the IT sector. The studies are in the computer science domain, which means the following 10 segments can be considered: (1) Software engineering, (2) Systems Engineering, (3) IT Management, (4) Information Storage and Retrieval, (5) Artificial Intelligence, (6) Programming languages, (7) Scientific Computing, (8) Computer Graphics, (9) Bioinformatics, and (10) Computer Architecture. Some specializations are well documented, for example the IEEE Computer Society established a Guide to Software Engineering Body of Knowledge (SWEBOK Guide [12]). One of the SWEBOK goals is to define educational curricula for undergraduates, graduates, and continuing education. There are 15 SWEBOK KAs (Knowledge Areas) distinguished.

A PhD student participates in core (basic) subjects that are the basis of well-justified solutions in the field of IT architecture. These subjects are of an agnostic nature as they are independent of technological solutions. The following 8 core courses supporting the acquisition of horizontal knowledge are proposed:

- Theoretical Foundation of Computer Science. The curriculum is based SWEBOK (Software Engineering Body of Knowledge—IEEE Computer Society) Chap. 13—Computing Foundations [12].
- System Design Engineering. The subject considers that a software system may be a part of other more general systems. The SEBoK (Systems Engineering Body of Knowledge Now Available—IEEE Computer Society [13]) could be a good reference.
- Advanced Data Base Systems. The subject considers topics: Distributed Databases, Distributed Transaction, Consensus Protocols, No SQL, NewSQL, Distributed Data Stores, Distributed Stream Processing, Alternative Data Storage & Model, Data Warehouses, Machine Learning Systems, OLTP/OLAP Hybrids, Crowdsourcing

- Software Architecture Design will be based in the first instance on an approach proposed by SEI (Software Engineering Institute of Carnegie Mellon University) using four main architecture development steps [14]: quality attribute workshop, attribute driven design, architecture documentation, and evaluation of architecture solutions.
- Mathematical Foundation for Computer Science. The subject is based on SWEBOk (Software Engineering Body of Knowledge—IEEE Computer Society) Chap. 14—Mathematical Foundations [12].
- Infrastructure Architecture. Infrastructure architecture has a strong influence on the quality of software systems and includes the following topics: Computer Networks, Performance Evaluation and Communication Networks, Broadband Networking Systems, Internetworking Architectures and Protocols, Networked Applications and Services, Access Management, Management (Deployment, Provisioning).
- Intelligent Systems. The following topics could be considered: Artificial Intelligence, Introduction to Cognitive Science, Computer Vision, Case-based Reasoning, Autonomous Robotics, Multi-Robot Systems, Computational Perception, Knowledge-Based AI, Machine Learning, Robot Intelligence: Planning, Natural Language Understanding, Philosophy of Cognition, Computational Data Analysis.
- IT Project Management Architecture. Knowledge of project management is important for IT architects. Projects can be realized based on classic (e.g. PMBOK), agile (e.g. Scrum) or mixed methodologies. The mentioned methodologies will be consistently and accurately described using ArchiMate IT architecture notation.

The specialist subjects are optional, which means students may choose subjects from a list. But to start preparation the following categories of elective subjects can be distinguished:

- New information technologies and novel IT solutions. In this case students will be well-prepared to deal with new technologies like cloud computing, big data, business intelligence, IoT (Internet of Things), social systems, etc.
- Advanced knowledge categories of architects' competency. In this case students will be prepared to gain knowledge of at least one advanced level IT architecture. ISAQB defines an Advanced-Level structure [15] as consisting of modules focusing on a particular core topic: architecture documentation, agile software architecture, architecture assessment, evolution improvement of software architectures, web architecture, soft skills for software architects, enterprise architecture management, service-oriented architecture, embedded systems.
- Current IT sector needs. Development is mostly based on adapting existing software solutions. For the IT sector the following tasks are important: legacy systems integration, business constraints, legal rules, time-to-market development, embedding in different contexts, integration with social environment, etc.
- PhD studies research agenda oriented courses (described in next section).

The aforementioned part-time studies are planned to start in the 2016/2017 academic year. The core subjects are already defined, however the categories of elective subjects are still under discussion with IT-sector companies and employees. After reviews and surveys the mapping which best meets the needs of the IT-Sector and universities will be selected.

6 Research Agenda

In Poland the performance measures of scientific entities are based on regulations from the Ministry of Science and Higher Education and consist of 4 basic criteria: scientific and/or creative achievements; scientific potentiality, tangible benefits of a scientific activity, intangible benefits of a scientific activity. Some publications explore the problem more accurately (e.g. [16]). According to these publications, PhD students should disseminate the results of their work by publishing or giving presentations at conferences and seminars. Writing scientific papers is necessary in order to obtain the formal approval of the proposed doctoral research and to gain acceptance for the defense of PhD thesis (disputation). The Ministry of Science and Higher Education publishes the number of credits awarded for publication in scientific journals.

The PhD student's scientific research is additionally supported by subjects and supplementary activities:

- Subjects that support scientific research intended to prepare PhD students for scientific research, communicate and commercialize its results, and prepare for doctoral exams;
- A PhD Seminar which provides students with an opportunity to gain skills with regard to obtaining information about research results and other scientific dissertations, to carry out their critical analysis (including the assessment of their own research) and to determine their usefulness and the possibility of practical implementation of their results.
- Cooperation with a coordinator and a supervisor. Scientific research is conducted under the aegis of a scientific coordinator and, when formal approval of the proposed doctoral research has been granted, under the aegis of the supervisor.

It is important that the PhD studies have a recognizable and visible research agenda. Designing a working agenda is not easy due to the facts that the available scientific resources are usually overloaded with other tasks, scientific research partially does not correspond to IT sector needs, and scientific knowledge on different IT re-search fields is dispersed.

The potential benefits of pursuing an intentional community-based development of a research agenda is presented in paper [17]. On the basis of the results of the aforementioned paper, a PhD agenda proposition is adapted by defining principles, goal and services (Fig. 2).

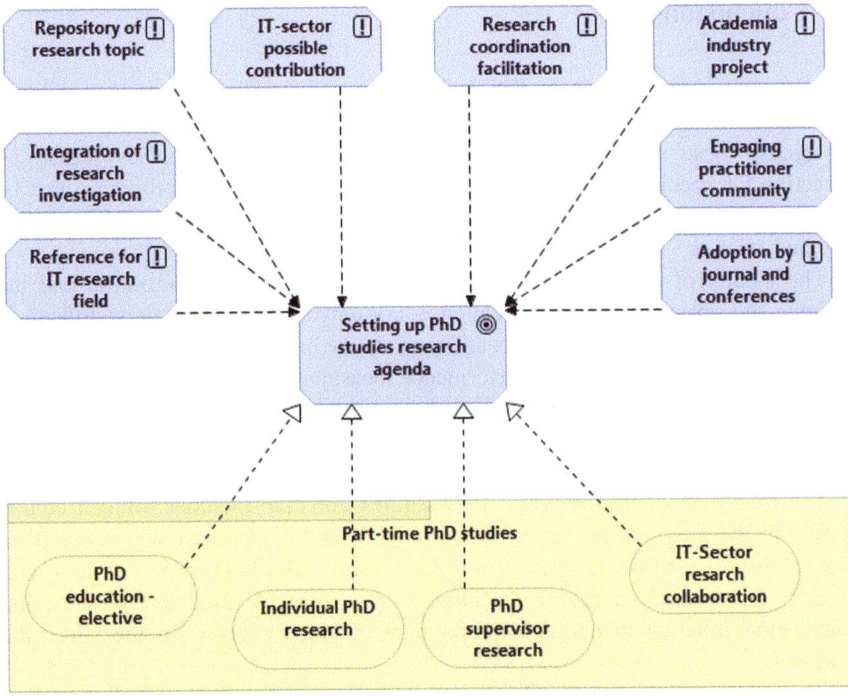

Fig. 2 Principles, goals and services proposed for PhD studies research agenda

It is interesting to define the specialized agenda which is of great importance to the IT sector. Let us consider an example of a requirement for PhD studies on trustworthy systems, for which leaders who will help protect our cyber systems' health are needed. The goal of a trustworthy systems program is to develop scholars with a wider scope than those who emerge from more traditional programs. In [18] the field of trustworthy systems is reviewed, and the state of trustworthy systems PhD programs in the US is examined. This example shows that specialized agenda of PhD studies in computer science could be important for some organizations in the IT sector.

Currently, the research agenda for the considered PhD studies is under development. Areas of the Part-time Doctoral Studies are related to solutions which are of current value to the IT sector: (1) Research areas related to systems and software and all aspects of software engineering which are associated with the issue of software allocation in hardware infrastructure; (2) Research areas related to the development of software. This concerns the improvement of software development processes; (3) Research areas related to corporate enterprise architectures, enterprise IT systems and software development within IT enterprises.

7 Conclusions

Developing university PhD studies that are best suited to the industry is a difficult but important task. Domain and non-domain features must be considered when de-signing these studies. The proposed 3rd level part-time studies should be attractive for specialists from the IT sector, therefore the study program must be well-organized, efficient, and based on current and important fields of IT technology and science.

The quality of studies can be achieved by proper study organization and the defining and setting of levels of excellence. For university staff it may be difficult to collaborate with many stakeholders and concentrate on values important to industry which coincide with university performance measures. For industry it may be hard to participate in research projects which do not produce results quickly. In general, the proposal should be beneficial both to universities and the IT sector. The proposed program has been constructed to be internationally recognized.

The road map of the proposed PhD studies initiative includes: official confirmation of the program and organization of PhD studies (the first step is complete and received a positive response from the Faculty); developing cooperation with industry representatives concerning interest areas, possible candidate profiles, and expectations relating to research; preparation of video courses for the e-learning platform.

It is of key importance for IT enterprises to employ highly-qualified specialists as IT architects and technicians play a strategic role in the development of IT systems. Competence in this field is directly related to the remuneration such specialists are offered. Part-time Doctoral Studies in Computer Science are the perfect solution for practitioners who wish to be perceived as highly-qualified specialists, and for that reason properly valued.

References

1. Bologna Declaration. http://en.wikipedia.org/wiki/Bologna_declaration
2. The Path Forward The Future of Graduate Education in the United States. Council of Graduate Schools and Educational Testing Service, (2010) http://www.fgereport.org/rsc/pdf/CFGE_report.pdf, p. 71
3. Howell Smith, M.C.: It's not what you think: A theory for understanding the lack of interest among domestic students in the engineering PhD. In: 41st ASEE/IEEE Frontiers in Education Conference (2011)
4. Smith, K.A., Streveler, R.A.: Special session—connecting and expanding the emerging engineering education research (EER) and engineering education innovation (EEI) communities. In: 41st ASEE/IEEE Frontiers in Education Conference (2011)
5. Bonnaud, O., Fremont, H., Thiriet, J.M., Yahoui, H.: PhD in electrical and information engineering in Europe: Towards a harmonization including LifeLong Learning. In: International Conference on Information Technology Based Higher Education and Training (ITHET) (2012)

6. Bonnaud, O., Fremont, H., Thiriet, J.M.: On the way of harmonization of PhD in Europe in electrical and information engineering: Status and recommendations. In: 23rd EAEEIE Annual Conference, p. 5 (2012)
7. Valdés, R.A., Moreno, J.C., García, F.J.S.E.: Educating engineering PhD students for a Global World. In: Global Engineering Education Conference (EDUCON) (2012)
8. Djuricic, A., Grady, H.M., Graham, W.G.: The information economy: Educational opportunities for industry-based professionals. In: IEEE International Professional Communication Conference (2008)
9. Jamróz, K., Pitulej, D., Werewka, J.: Adapting enterprise architecture at a software development company and the resultant benefits. In: Avgeriou P., Zdun U. (eds.) ECSA 2014, LNCS 8627, pp. 170–185 (2014)
10. IT Architecture Academy. www.it-architecture.agh.edu.pl
11. Turek, M., Werewka, J.: Motivation Modeling and Metrics Evaluation of IT Architect Certification Programs, BDAS15, Communications in Computer and Information Science. Springer, London (2015)
12. Guide to the Software Engineering Body of Knowledge, Version 3.0, SWEBOK®, A Project of the IEEE Computer Society. In: Bourque P., (Dick) R.E. (eds.) Fairley, p. 335 (2014)
13. Guide to the Systems Engineering Body of Knowledge (SEBoK), v1.3.2. IEEE, pp. 976 (2015)
14. Bass, L., Clements, P., Kazman, R.: Software Architecture in Practice (3rd edn) (SEI Series in Software Engineering) (2012)
15. Acquiring skills through the Advanced-Level scheme, iSAQB (International Software Architecture Qualification Board), http://www.isaqb.org/certifications/advanced-level
16. Koczkodaj, W., Kułakowski, K., Ligęza, A.: On the quality evaluation of scientific entities in Poland supported by consistency-driven pairwise comparisons method. Scientometrics (2014)
17. Agresti, W.W.: Advancing the IT Research Agenda, RIIT'13, pp. 11–16 (2013)
18. Yasinsac, A.: Cynthia Irvine. Help! Is there a trustworthy-systems doctor in the house? IEEE Secur. Priv. 73–77 (2013)

Multi-project Scrum Methodology for Projects Using Software Product Lines

Michał Turek and Jan Werewka

Abstract The article describes a new solution that adapts agile Scrum methodology to function in multi-project environments. The goal of this project is a common software framework which is simultaneously developed by multiple Scrum teams. We focus on a software framework case which is developed for many customers concurrently using a software product line characterized by reusable common functionality. The solution provides a detailed software cost reckoning for commonly developed software components which enables companies to settle software production costs with customers.

Keywords Software product lines · Software development · Agile · Scrum · UAV

1 Introduction

The basic assumptions of Scrum methodology derive from the studies of professors Hirotaka Takeuchi and Ikujiro Nonaka. In their paper [1], which is widely regarded as the source of inspiration for Scrum, the authors presented the so-called holistic approach that should be very effective for developing innovative products.

The views of H. Takeuchi and I. Nonaka were a direct inspiration for Ken Schwaber, Jeff Sutherland, and Mike Beedle to formulate the Scrum methodology containing many references to the principles of Agile Manifesto and the features of a holistic process. At the 1995 OOPSLA conference, Ken Schwaber presented a

M. Turek (✉) · J. Werewka
Department of Applied Computer Science, AGH University of Science and Technology, al. Mickiewicza 30, 30-059 Kraków, Poland
e-mail: mitu@agh.edu.pl

J. Werewka
e-mail: Jan.Werewka@agh.edu.pl

© Springer International Publishing Switzerland 2016

J. Świątek et al. (eds.), *Information Systems Architecture and Technology: Proceedings of 36th International Conference on Information Systems Architecture and Technology – ISAT 2015 – Part III*, Advances in Intelligent Systems and Computing 431, DOI 10.1007/978-3-319-28564-1_16

formal definition of the Scrum methodology [2]. Mike Cohn [3] adopted agile project development from the perspective of product ownership.

Book [4] presents a systematic description of Scrum, which is the most popular agile methodology that combines project management and product development. The authors propose a meta-model and a set of basic concepts to describe different methodologies of this type. The basic postulate of agile methodologies is to satisfy the customer's needs. The incremental approach and frequent exchange of information about requirements during project development ensures better predictability and risk control.

Software product line engineering (SPLE) has received a lot of attention with regards to improving software development productivity. Clements and Northrop [5] define a software product line (SPL) as a set of software-intensive systems (products) that share a common, managed set of features which satisfy the specific needs of a particular market segment or mission and are developed from a common set of core assets in a prescribed way. A software product line enables the creation of a number of similar products by selecting and composing reusable software artifacts. The key aspect of developing a SPL is to capture the commonalities and variabilities of soft-ware product line members.

Software product lines which have many common features are introduced in companies which work for multiple customers. This is a very common situation which is currently especially important for small and medium size software companies. The main SPL is developed as a framework and is sold under license. This paper considers a special case in which a customer buys not only a framework license, but also orders a series of software modifications. These modifications can in some cases affect the framework, thereby forcing or simply allowing a company to enhance framework functionality (usually to meet a particular customer's requirements). In some cases, when the utility of these enhancements is high, they can be left within a common framework for future use and re-sale. This article discusses situations in which a customer pays for ordered software, but the software itself can be re-used as a part of a framework and therefore should be taken under consideration in the cost-settlement field. This software evolution pattern is very common in small companies which provide, for example, long-term highly customized Internet service products of some kind (Fig. 1).

Another issue considered is the framework branching problem. Continuous frame-work modifications from many sources can easily cause software line branches in framework development. There is no agile methodology that can successfully deal with this situation.

There is some discussion about contracts with agile teams. The goal presented in paper [6] is to meet the customer's needs while maintaining commitment to agile practices. The different types of contracts are discussed there: Time and Materials (T&M), Hybrid of T&M and Fixed Price, Fixed Scope, and Fixed Schedule contracts supporting agile development. Paper [7] presents experience gained from conducting collaborative contracts which supported agile development and encouraged efficient collaboration between customer and supplier.

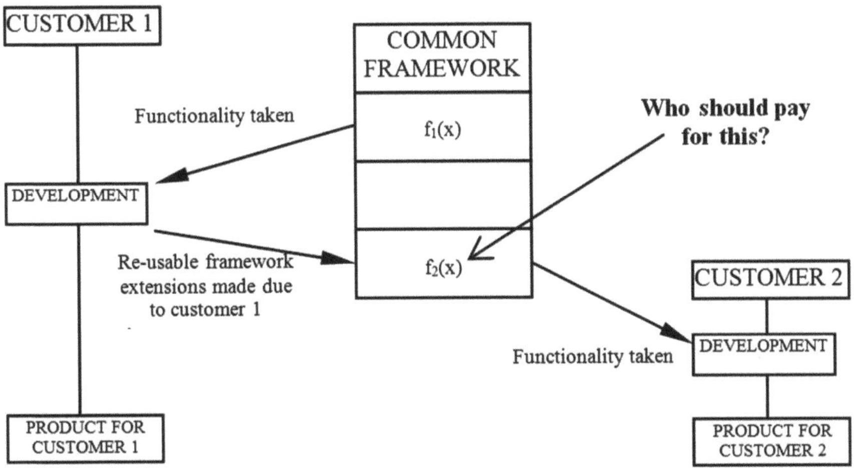

Fig. 1 Common framework with re-usable content

Working with different customers on similar projects can result in inconvenient situations in which some customers are not clear about what they are paying for. The goal of the work described in this article is also to establish effort and cost effective methods for common software framework management using a Scrum agile approach. For this situation a story point trading technique will be used which allows the estimation and negotiation of the cost of a product which is developed in non-exclusive mode (meaning it will also be re-used by an SDC (Software Development Company)).

2 Adapting Scrum Methodology for a Multi-project Environment

It is common knowledge that Scrum agile methodology is used when requirements are unstable and a job must be divided into small iterations characterized by consequent product business value increase, while reducing the risk of software development. The methodology defines two special registers called Product backlog (aggregates tasks to be done), and Sprint backlog (jobs to be done in the current iteration—called Sprint). The methodology also assumes precise effort settlement with a scalar velocity measurement (usually using a story-points scale or similar). The product backlog is reduced after each iteration until a final business-worthy product version is released. Then the development process may be continued in order to achieve another product version which is bound to another product's backlog entries. In a classical Scrum approach there are many types of Sprint cost settlements. The easiest and most popular way is to settle each sprint as the fixed amount of

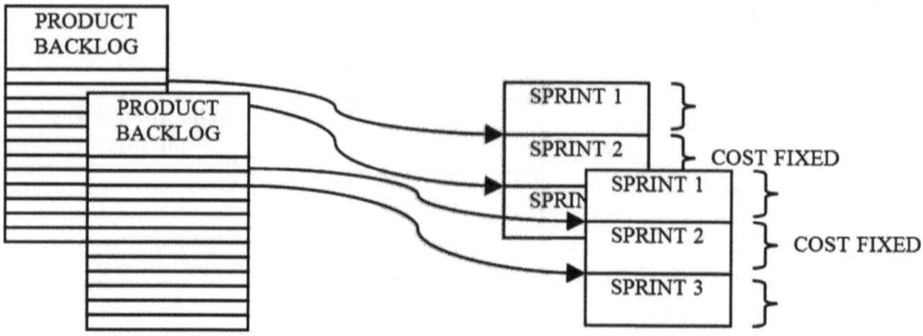

Fig. 2 A typical cost estimation in a Scrum multi-project environment with isolated Scrum teams

men-hours spent on it (sprint has a fixed running time expressed in working days) (Fig. 2).

A multi project environment assumes that Scrum teams in a company are working on different products (or product components). In such an environment there are human resource collisions, module integration problems, and other well-known issues for which solutions exist. A special situation arises when some parts of a product being developed can be re-used by many Scrum teams at the same time (same iteration) or at a later time (subsequent). The common part of the software must be man-aged somehow, but no methodology yet exists for this scenario. The situation be-comes slightly more complicated when there are many similar ongoing projects in a company due to the fact that there many Scrum teams working at the same time. People in these teams will be working on different projects but with common resources involved. Methods for resource sharing in such cases are well-known so will not be considered now. But an interesting question arises: how can a product increment be re-used in real time and how can product development cost be calculated using story-points or other effort accounting systems?

3 Developing Common Software Framework in a Scrum Multi-project Environment

In the proposed approach multiple product backlogs will be used. One for a special team developing common framework software and a few others for Scrum teams working for different customers on products which are extensions of a common framework. Each product backlog will contain a sorted set of Stories (as it should be in Scrum classical methodology) (Fig. 3).

Decomposed items in the product backlog will be estimated in a special way by not only generating an effort estimation measured in story points (usual Scrum procedure), but also by making a relation that expresses the level of interest in a

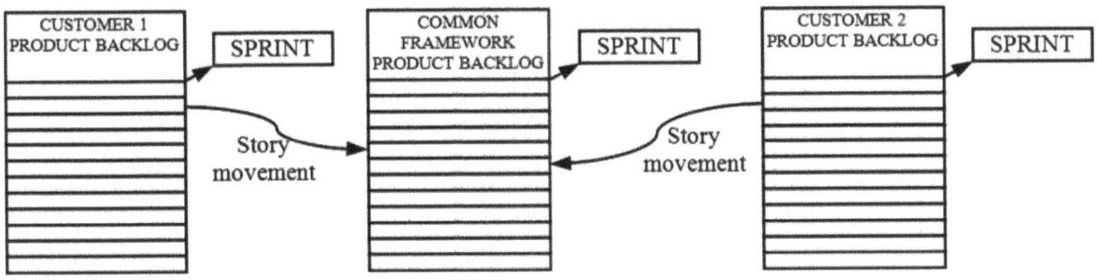

Fig. 3 Multiple product backlogs with a common framework product backlog

story for Scrum teams in an SDC. This factor's value will be negotiated between the Product owner and the Scrum team during the estimation process (the Scrum team here represents SDC strategy). If the SDC finds some story worth putting in a common frame-work (meaning: "it could be used again"), the Scrum team starts negotiations with the Product owner. There are two goals for such negotiations:

- A factor value agreement. The factor itself is just a value in the range 0 to 1 which expresses the interest of product owners in some story. A factor value of 1 means the functionality described by the story is developed exclusively for the product owner. Lower values mean the SDC is also interested in placing the story's product increment in a common framework.
- Choice of Scrum team activity. A Scrum team may be assigned to either developing customer functionality only, or working on common software framework functionality. At this point negotiations could result in:
  - A story being moved to a common framework Product backlog, further decomposed and processed by a common software framework Scrum Team,
  - A story remaining in the current backlog, but, due to its reusability, the product owner not paying the full cost for the story points used.

In many situations a story will indeed be moved to a common software framework product backlog. This is preferable for the following reasons.

- Due to familiar technology, a common framework Scrum team can more efficiently lead framework development and achieve significantly better product quality as the product architecture is also familiar,
- In some cases there is a chance of immediate product increment reuse, when at least two customer's projects concurrently need it.

A story is ready for development (DoR—Definition of Ready) when it is clear and the team has all the information needed to proceed with the story. The DoR metric should be accompanied by a Scrum INVEST or similar rule. Bill Wake used the acronym "INVEST" [8] to define the characteristics of a good story (I—Independent, N—Negotiable, V—Valuable, E—Estimable, S—Small, T—Testable). The proposed approach allows a story to be placed in a sprint backlog only if it is small enough to say who should accomplish it (a common software framework Scrum team or a dedicated one).

Common software framework Scrum team planning meetings will be held in the presence of all product owners who send their stories (hi) to the common software framework product backlog (PBL), thus making further story decomposition possible.

$$\mathrm{PBL} = \{\mathrm{h}_1, \ldots \mathrm{h}_i, \ldots \mathrm{h}_n\} \quad (1)$$

If some stories are moved to a common software framework product backlog, the product owner pays for the stories done by a common software framework Scrum team in accordance with the formula:

$$Domain\,Team\,Payment(h_i) = C_{SDT}\frac{DTSP(h_i)}{V_{DT}}f(h_i) \quad (2)$$

.

- C_{SDT}—Cost of a sprint for the domain team
- $DTSP(h_i)$—is the amount of story points used (assigned to a story during last estimation) by domain team for history h_i,
- V_{DT}—is a common framework (domain) Scrum team's velocity for a sprint (expressed in story points),
- $F(h_i)$—is a reducing factor for a story payment.

There could always be a situation in which two or more customers (product owners) are interested in developing the same functionality. In such a case, the cost can be split using common software framework (domain) team.

When a story is of interest to the SDC but is not moved to a framework team, the Product owner receives a discount according to the formula:

$$Application\,Team\,Discount(h_i) = C_{SAT}\frac{ATSP(h_i)}{V_{AT}}(1 - f(h_i)) \quad (3)$$

- C_{SAT}—Cost of a sprint for the application team
- $ATSP(h_i)$—is the amount of story points assigned during last estimation by application team for history h_i,
- V_{AT}—is an application Scrum team's velocity for a sprint (expressed in story points),

A story-point factor negotiation technique allows effort values to be estimated for newly developed common software framework components with a fair division of development costs.

4 Scrum Approach to Software Product Lines

When developing an SPL it is necessary to solve the following problems [9]: (1) software development for reuse in domain engineering; (2) platform definition, which consists of defining a basic product line architecture for which common and variant features are planned and defined; (3) development with reuse in application engineering, meaning that using reusable assets inside the product line architecture ensures that a specific product is obtained.

In the literature some examples are given for adoption of the SPL approach to agile methods and feature relation which may be used for the development of effort calculation. The research [10] explores the interoperability and complementarities of the lean and agile approaches in combination with a software product line engineering approach. This is based on: (1) complementing architectural issues in large scale lean and agile development and (2) providing methodological guidance to make product line engineering more agile and efficient. In [11] a practical method for variation management with respect to pair-wise feature interactions is presented. In [12] an architectural approach is presented that supports multiple concerns. The approach presented in this paper comprises two complementary parts: (1) an update viewpoint that defines the conventions for constructing and using architecture views to deal with multiple update concerns; (2) a supporting framework that provides an extensible infrastructure supporting integrators of an SPL.

In [13] the application of Software Product Line (SPL) technology in the video game domain is explored by exploiting differences in various video game platforms in order to design a variable component-based software product line architecture for a multiplatform video game. The approach consists of constructing a feature dependency model for describing variability in multiplatform video games. Existing approaches have several limitations, such as a lack of quantitative measurements or the need for existing valid products and intensive human effort for the assessment.

In our solution it is important to find the relation between the customer's products and the produced software components according to the software line principle (Fig. 4). Customer 1 is interested in product 1, customer 2 uses product 2 and so on. Each of the products is characterized by features which might be common for different products and are elements of one or several Scrum product backlogs. The teams develop software by delivering core, variable or custom software components.

Table 1 shows the relation between products and components and may be used in calculating the costs of Scrum multi-project contracts. From the table it can be deter-mined in which products a given component will be included. In this way an effort and cost path can be determined.

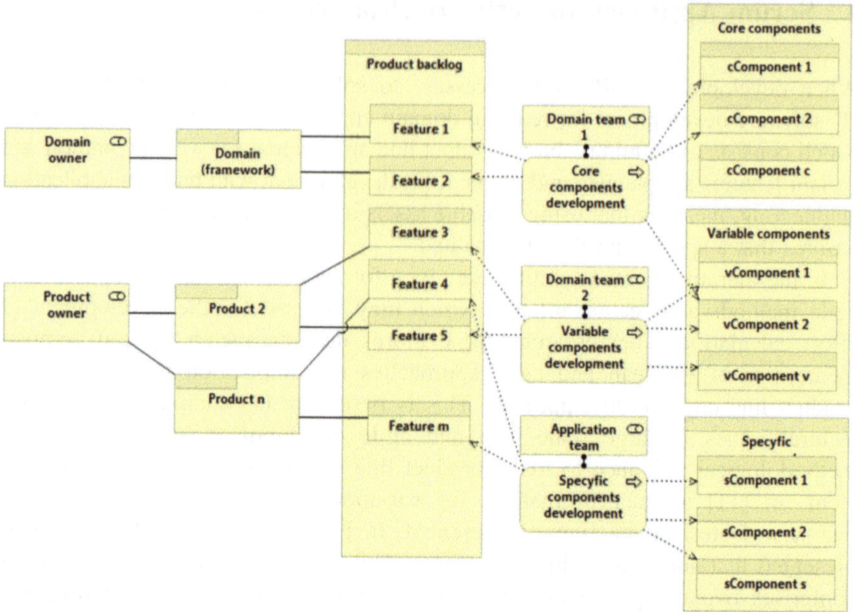

Fig. 4 Scrum software product line structure

Table 1 Relation between products, features and components

	Products					Components								
		P1	P2	P3		C1	C2	C3	C4	C5	C6	C7	C8	C9
Features	F1		X			X		X		X	X	X		
	F2	X	X	X			X		X			X		X
	F3		X	X			X						X	X
	F4	X				X	X	X						
	F5		X	X					X					
	F6	X						X						
	F7	X	X							X				

5 Development of a UAV System Based on SPL

There many success stories in developing software using the software product lines approach. An interesting case is the development of a UAV (Unmanned Aerial Vehicle) system. Development of such innovative systems causes changes of product line at both application and domain level. The UAV system structure may consist of the following subsystems (Fig. 5):

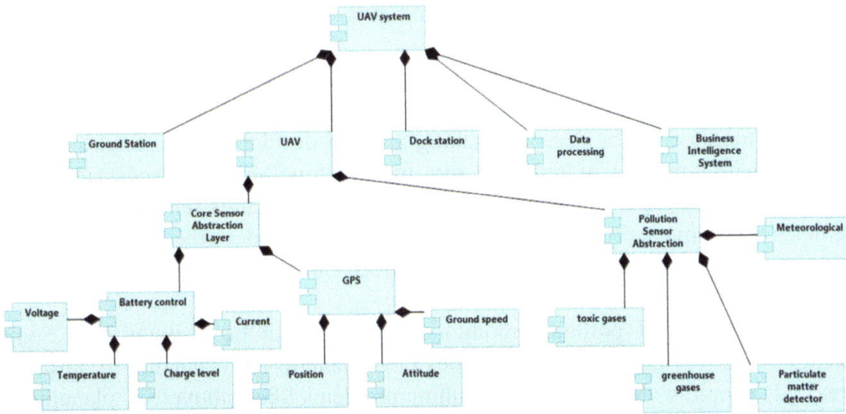

Fig. 5 UAV system selected components

1. UAV equipped with
   (a) Flight Control System, responsible for transporting a pay-load.
   (b) Mission System, the part of the system responsible for the pay-load.
2. Ground Control station, which is responsible for UAV remote control, landing and take offs, route planning, etc.
3. Docking station, used for UAV battery regeneration and as an interface for the Ground Control Station and the Data Processing System.
4. Data Processing system, used for processing on-line and off-line mission data. The system is responsible for appropriate data storage.
5. Business Intelligence System, used to investigate dependencies and reasoning of the current set of missions

Considering the aforementioned subsystems, it can be said that candidates in an SPL for core elements are subsystems (1) a, (2), (3), (4) and the candidates for variable or custom subsystems are (1) b, (5).

When developing SPL architecture it is important to know the different types of existing architectures. In [14] the architecture description for a small UAV is given. The system is composed of a sensor subsystem, a navigation and control subsystem, a wireless communication subsystem, an avionics subsystem, a power management subsystem, and an actuator subsystem. In [15] a HAMSTER data communication architecture is proposed in order to help developers of unmanned aircrafts efficiently implement communications in UAS by considering internal and external communications. In [16] a Flight Control System Gateway is designed to facilitate exploitation of data obtained from the autopilot. The gateway provides a hardware-independent interface that isolates payload components from the autopilot specificities, thus eliminating dependencies on a particular autopilot solution. These subsystems are integrated into an overall avionics solution which is based on a publish/subscribe architecture pattern and a LAN based distributed hardware architecture.

A UAV may have very broad applications. An atmospheric pollution monitor-ing system was chosen as an example. The goal of such a system is on-line pollution monitoring for decision making and the development of environmental policies, with the goal of reducing the impact of pollution on ecosystems and human health. In paper [17] a method is described which uses sensors mounted on vehicles to automatically detect air pollution and fog. The described system consists of sensors which acquire their primary data from cameras and Light Detection and Recognition (LIDAR) instruments.

Different users of pollution monitoring systems can be distinguished. Some examples are: (1) Individuals or families (e.g. planning an excursion in an urban area); (2) People with allergies or breathing problems like asthma (e.g. planning to go outside); (3) Schools (e.g. planning different sports activities outside for children); (4) Early warning systems for municipal operation; (5) Urban planning used for air pollution reduction. There are different types of pollution sensors which can be wearable, mounted on vehicles or UAVs, or installed in fixed positions (buildings, weather stations, etc.). The specialization of pollution sensors may differ significantly: (1) Personal environmental sensors, which are wearable sensors connected to or integrated with smart phones; (2) Pollution sensor stations installed on fixed points or on vehicles measuring air pollutants; (3) LIDAR (Light Detection and Recognition) used to detect small concentrations of air pollutants.

In Fig. 5 a segment of UAV systems is presented which shows the structure of sensors for flight control and mission systems. The mission could be to obtain e.g. Air Quality Indexes [18] to enhance environmental awareness. The presented discussion aims only to be an overview of possible pollution system monitoring solutions, and is not intended to demonstrate the full picture.

6 Conclusions

In the paper a solution is presented for dealing with an agile approach in a multi-project environment which produces software with similar features for different customers. The proposed solution may be applied to small companies which produce their own software product lines. The development process is still based on classical Scrum story sorting procedures, but allows story migration between different product backlogs. New common software framework component development is estimated with a fair cost division between all interested parties. It is possible to identify story similarity not only within one product backlog, but between product backlogs led by different product owners. A UAV systems case study shows a possible approach to software product lines in an agile environment.

References

1. Takeuchi, H., Nonaka, I.: The new new product development game. Harvard Bus. Rev. **64**(1), 137–146 (1986)
2. Schwaber, K., Sutherland, J.: The scrum guide™, the definitive guide to scrum: The rules of the game. http://www.scrum.org/scrumguides (2013)
3. Cohn, M.: Agile estimating and planning. Pearson Education, UK (2006)
4. Werewka, J., Nalepa, G.J., Turek, M., Włodarek, T., Bobek, S., Kaczor, K.: Project management in IT company. In: Project and Software Development Process Management (in Polish), Vol. 3, p. 164. AGH University of Science and Technology Press, Krakow (2012)
5. Clements, P., Northrop, L.: Software Product Lines. Addison-Wesley, Reading (2001)
6. Franklin, T.: Adventures in Agile Contracting: Evolving from Time and Materials to Fixed Price. Fixed Scope Contracts, IEEE (2008)
7. Throup, L., Jensen, B.: Collaborative agile contracts. In: Agile Conference, pp. 196–200 (2009)
8. Wake, B.: Invest in good stories, and smart tasks. http://xp123.com/articles/invest-in-good-stories-and-smart-tasks/ (2003)
9. Zuppiroli, S., Ciancarini, P., Gabbrielli, M.: A role-playing game for a software engineering lab: Developing a product line. In: 25th IEEE Conference on Software Engineering Education and Training, pp. 13–22 (2012)
10. Blau, B., Hildenbrand, T.: Product line engineering in large-scale lean and agile software product development environments. Towards a hybrid approach to decentral control and managed reuse. In: Sixth International Conference on Availability, Reliability and Security, pp. 404–408
11. Kato, S., Yamaguchi, N.: Variation management for software product lines with cumulative coverage of feature interactions. In: 15th International Soft Product Line Conference, pp. 140–149 (2011)
12. Weyns, D., Michalik, B., Helleboogh, A., Boucké, N.: An architectural approach to support online updates of software product lines. In: WICSA, pp. 204–211 (2011)
13. Albassam, E., Gomaa, H.: Applying software product lines to multiplatform video games. GAS (2013)
14. Chong, M., Chuntao, L.: Design of flight control software for small unmanned aerial vehicle based on VxWorks. In: Proceedings of 2014 IEEE Chinese Guidance, Navigation and Control Conference (2014)
15. Pigatto, D.F., Gonçalves, L., Pinto, A.S., Roberto, G.F., Filho, J.F., Branco, K.R.: HAMSTER —healthy, mobility and security-based data communication architecture for unmanned aircraft systems. In: International Conference on Unmanned Aircraft Systems (ICUAS) (2014)
16. Santamaria, E., Royo, P., Lopez, J., Barrado, C., Pastor, E., Prat, X.: Increasing UAV capabilities through autopilot and flight plan abstraction. In: IEEE Conference on Digital Avionics Systems Conference (2007)
17. Sallis, P., Dannheim, C., Icking, C., Maeder, M.: Air pollution and fog detection through vehicular sensors. In: Modelling Symposium (AMS), 8th Asia, pp. 181–186 (2014)
18. Penza, M., Suriano, D., Villani, M., Spinelle, L., Gerboles, M.: Towards air quality indices in smart cities by calibrated low-cost sensors applied 1to networks. In: IEEE (2014)

Test Scenarios and Validation Results of POP Methodology

Grzegorz Kołaczek, Paweł Świątek, Jerzy Świątek, Krzysztof Brzostowski, Adam Grzech and Arkadiusz Sławek

Abstract The paper presents the results of quantitative and qualitative assessment of effectiveness of POP Methodology (a selection of agile methods of analysis, planning and optimization of business process management information systems, developed by a team of researchers at PWr). In particular, the POP Methodology evaluation metrics are proposed, focusing mainly on the process of software product development and the product quality assessment in various stages of the process. Factors such as duration, experience and size of a development team are considered. The proposed metrics are discussed and illustrated based on a few selected examples.

Keywords Software development · Process metrics · Evaluation

G. Kołaczek (✉) · P. Świątek · J. Świątek · K. Brzostowski · A. Grzech · A. Sławek
Chair of Computer Science, Wrocław University of Technology,
Wybrzeże Wyspiańskiego 27, 50-370 Wrocław, Poland
e-mail: grzegorz.kolaczek@pwr.edu.pl

P. Świątek
e-mail: pawel.swiatek@pwr.edu.pl

J. Świątek
e-mail: jerzy.swiatek@pwr.edu.pl

K. Brzostowski
e-mail: krzysztof.brzostowski@pwr.edu.pl

A. Grzech
e-mail: adam.grzech@pwr.edu.pl

A. Sławek
e-mail: arkadiusz.slawek@pwr.edu.pl

© Springer International Publishing Switzerland 2016

J. Świątek et al. (eds.), *Information Systems Architecture and Technology: Proceedings of 36th International Conference on Information Systems Architecture and Technology – ISAT 2015 – Part III*, Advances in Intelligent Systems and Computing 431, DOI 10.1007/978-3-319-28564-1_17

1 Introduction

The goal of the research presented in this paper is quantitative and qualitative assessment of effectiveness of the POP Methodology [1–3]. The POP Methodology facilitates selection of methods of analysis, planning and optimization of business processes managed in domain-specific enterprise resource management information systems. The results of the assessment will be used to plan and implement improvements of some elements of the POP Methodology in its next versions.

In order to test and validate the POP Methodology and tools supporting the Methodology, implemented in the POP Platform [1, 4] a number of test scenarios have been developed based on examples of real business processes in transportation systems. The next part of the paper describes evaluation methods used to assess the POP Methodology and how they were executed in a dedicated and preconfigured validation environment. The results of the evaluation focus mainly on the development process defined by the POP Methodology and on quality assessment of software products developed using this process.

The POP Methodology proposes a complete procedure to be followed by development teams specializing in solving optimization problems. It defines a series of steps, from the very beginning of business analysis, through the structured specification of an optimization problem, to the final delivery of a prototype solution—a software implementation of an algorithm solving an identified optimization problem. The POP Methodology also defines a set of technics and software tools which support the procedure [1, 4]. These elements are as follows:

- A tool to assist work of an analyst
- A method for identifying optimization needs,
- A collection of dictionaries and corresponding development tools,
- A repository of optimization problems models,
- A method for selecting or generating algorithms solving optimization problems,
- A repository of algorithms solving optimization problems.

2 Related Works

There are more than fifty software development methods and techniques which are currently in use and many more hybrids combining elements of multiple approaches into a final one. The set of the development methods starts with the traditional waterfall approach, goes through some proprietary methodologies and ends with various types of agile frameworks. Some examples (in no particular order) are: the Rational Unified Process (RUP), the Team Software Process (TSP), V-Model development, Microsoft Solutions Framework, the Structured Analysis and Design Technique (SADT), Evolutionary Development (EVO), Extreme Programming (XP), PRINCE2, Merise, model-based development, and many more [5–8]. All these methods and techniques vary in complexity, scope, application and other aspects. Even if only a handful of them are compared and a few common elements are being investigated, that is still quite a significant amount of information to be collected and analyzed [9]. This is why extensive studies on comparison of different software methodologies are not very frequent [10, 11]. In result, quite frequently the choice of software development methodology has no rational justification and is not based on making a technical decision [13]. Many companies do not even attempt to evaluate alternative methods, but merely adopt the most popular, which today constitute the many faces of agile [10]. Nevertheless, over the years some objective measures have been defined and applied to assess the effectiveness of methodologies and (more often) their execution in software development projects.

The most frequent attempt to compare processes of software development is performed using three major domains [11]:

- Speed: Development schedules, effort,
- Quality: Software quality in terms of delivered defects,
- Economics: Total Cost of Ownership (TCO) and Cost of Quality (COQ).

There are sets of measures defined for all of the above domains. These software development metrics can be classified into three groups:

- Product metrics,
- Process metrics,
- Project metrics.

The product metrics describe characteristics of a product such as size, complexity, design features, performance, and quality level. The process metrics can be used to improve software development and maintenance. Examples include: effectiveness of defect removal during development, pattern of testing defect arrival and response time of the fix development process. The project metrics describe project characteristics and execution. Examples include: number of software developers, staffing pattern over software lifecycle, cost, schedule, and productivity.

Some metrics belong to multiple categories. Software quality metrics are a subset of software development metrics that focus on the quality aspects of the product, process, and project. In general, the software quality metrics are more closely associated with process and product metrics rather than with project metrics. Nonetheless, project parameters such as the number of developers and their skill levels, the schedule, the size, and the organization structure certainly affect the quality of the product.

Product quality metrics can be divided further into end-product quality metrics and in-process quality metrics. One can view quality from the entire software lifecycle perspective and, in this regard, one should include metrics that measure the quality level of maintenance process as another category of software quality metrics [5, 6, 9].

In the next chapters selected metrics are presented and discussed in two main groups of software quality aspects: product quality and process quality.

3 The POP Methodology Process Metrics and Process Quality Evaluation

The POP Methodology has been validated and analyzed using quantitative and qualitative measures. A set of process metrics has been proposed based on standards and best practices dedicated to the evaluation of software development process quality [3, 12] and included the following:

1. Time and resource consumption (quantitative measures: time, cost, number of experts and level of expertise required)
   (a) Time required for a task completion (measure: number of days)
      (i) Time required for business analysis—corresponds to the "Task description" step of the POP Methodology
      (ii) Time required for consultations with a customer—corresponds to the "Task description" step of the POP Methodology
      (iii) Time required for problem modeling using Process Optimization Platform tools—corresponds to the "Problem modeling" step of the POP Methodology
      (iv) Time required for a new library preparation—corresponds to the "Software library" step of the POP Methodology
      (v) Time required for a new algorithm development—corresponds to the "New algorithm" step of the POP Methodology
      (vi) Time required for a test dataset preparation—corresponds to the "Model selection" and "Data import" steps of the POP Methodology
      (vii) Time required for testing software library and solution algorithm-corresponds to the "New algorithm" step of the POP Methodology
   (b) Size of a team (number of experts)
   (c) Team experience (average experience of the team members: small, medium, high)
   (d) Total costs of the problem analysis and prototype development.

3.1 Preliminary Results of the POP Methodology Process Quality Evaluation

Evaluation of process defined by the POP Methodology was performed based on a number of case studies. The data relevant to the process metrics have been collected during analysis and solution development for optimization problems identified in real business processes occurring in four selected organizations. Three of datasets are related to the path no. 4 of the POP Methodology [2]. It means that in three out

Table 1 Total time required for analysis and algorithm development

Organization	Methodology path	Total time person/day
ORG1	Path 4	742
ORG2	Path 4	854
ORG3	Path 4	633
ORG4	Path 1	232

of four cases a complete analysis, modeling and solution development had been performed. Those steps have been performed both with and without the use of tools included in the POP Platform. The last dataset represents path no. 1 of the POP Methodology. In this case, results of previously completed analysis, modeling and algorithm implementation had been reused.

Results of the evaluation show that the most time consuming elements of the process were the "Task description" and "Problem modeling" steps of the POP Methodology. It confirmed that analysis of different types of organizations requires a significant time to prepare a relevant model of the optimization problem—especially when performed from scratch. However, the results proved that the POP Methodology and related tools can speed up the process of analysis, modeling and solution development not only in scenarios when the new problem is similar to one of previously analyzed cases and when previously developed solutions can be directly reused, but also when the new problem is original but belongs to the same domain. The use of tools included in the POP Platform (domain dictionaries, model and algorithm repositories and their corresponding editors) plays a key role in facilitating the "Task description" and "Problem modeling" steps and in result—the whole process (Table 1).

4 The POP Methodology Product Metrics and Product Quality Evaluation

The POP Methodology defines an agile method of developing software solutions for optimization problems [2]. The iterative nature of the defined development process imposes the need to assess product quality both from the end-product and in-process perspective. The product quality has been evaluated at the end of every iteration of the development process.

Evaluation of product quality in the POP Methodology context involves assessing the completeness of an optimization problem description and assessing the compliance of functionality offered by the final software product with requirements defined by users in a particular business environment. The following types of measures define the POP Methodology product quality evaluation:

- Adequacy of optimization problem identification,
- Completeness of optimization problem identification,

- Identification of optimization problem inputs,
- Identification of optimization problem outputs,
- Identification of optimization criteria,
- Identification of optimization problem constraints.

Based on the above assumptions a set of product metrics has been proposed and included the following:

2. The quality of analysis
   (a) number of identified optimization problems (numerical value)
   (b) completeness of identified optimization problems (the number of identified problems/the number of problems to be solved by the software solution)
   (c) adequacy of mapping between the set of identified decision making tasks in an organization and the set of identified optimization problems
      (i) adequacy of inputs (number of adequate inputs/total number of inputs)
      (ii) adequacy of outputs (number of adequate outputs/total number of outputs)
      (iii) adequacy of optimization criteria (number of adequate optimization criteria/total number of optimization criteria)
      (iv) adequacy of constraints (number of adequate constraints/total number of constraints)

The following method of calculating the above product metrics have been proposed (Table 2).

4.1 Preliminary Results of the POP Methodology Product Quality Evaluation

The POP Methodology product quality evaluation results, presented in this chapter, have been based on data obtained during one of the case-studies mentioned in Chap. 3. In the considered case, the process, facilitated by the POP Methodology and POP Platform tools, led to creation of a product—a prototype software solution of a real optimization problem, identified in an organization operating in the transportation domain. There were four iterations before the final product was completed, each iteration being a complete cycle of software development, which started with the problem definition and ended with the prototype presentation to the client.

The product quality evaluation has been performed at the end of each iteration in order to assess both the intermediate results and the final product. Each time the values of corresponding product metrics have been calculated. The figures presented below illustrate changes in the results adequacy level (Fig. 1), completeness level of identified optimization problem inputs (Fig. 2) and algorithms reusability level (Fig. 3).

Table 2 Product quality metrics

Goal	Question	Metrics	Interpretation
Verification of results adequacy	To what extent identified optimization problems address requirements?	X = 1—A/B A—number of wrongly identified optimization problems B—total number of identified optimization problems	0 <= X <=1 X is closer to 1 then adequacy is higher
Verification of completeness of inputs (outputs/criteria/constraints)	Is the list of identified inputs to optimization problems complete? (outputs/criteria/constraints)	X = 1—A/B A—number of missing inputs B—total number of inputs (outputs/criteria/constraints)	0 <= X <=1 X is closer to 1 then completeness is higher
Verification of the reusability of the model	To what extent the created model of optimization problem is reusable?	X = 1—A/B A—number of newly defined models B—number of models used	0 <= X <=1 X is closer to 1 then reusability is higher
Verification of the reusability of the classes	To what extent the created classes of optimization problem are reusable?	X = 1—A/B A—number of newly defined classes B—number of classes used	0 <= X <=1 X is closer to 1 then reusability is higher
Verification of the reusability of the algorithms	To what extent the created algorithms solving optimization problem are reusable?	X = 1—A/B A—number of newly defined algorithms B—number of algorithms used	0 <= X <=1 X is closer to 1 then reusability is higher

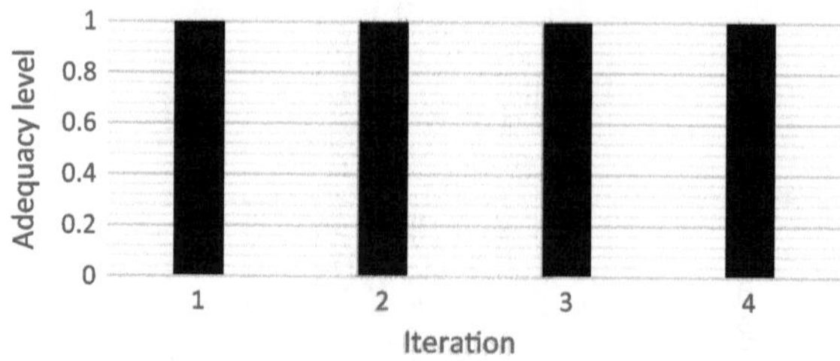

Fig. 1 Value of results adequacy

Fig. 2 Completeness level of identified optimization problems inputs

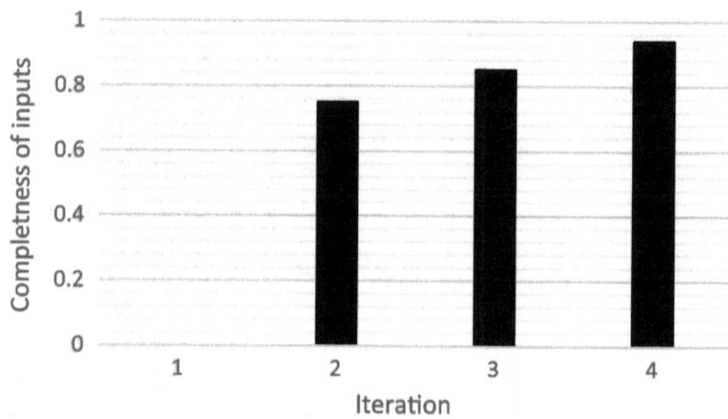

Fig. 3 Algorithms reusability level

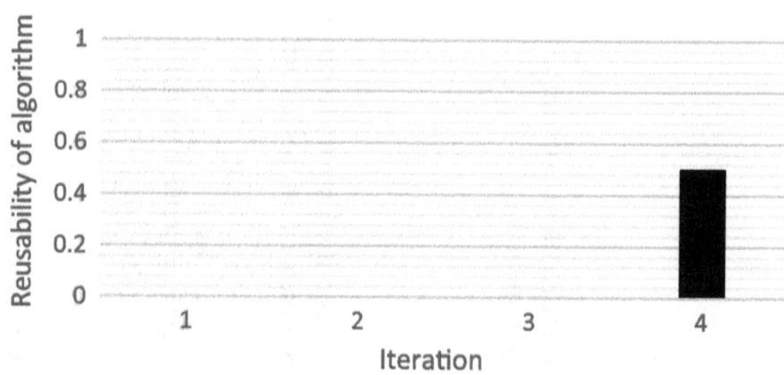

The obtained results (Fig. 1) show that adequacy of optimization problem identification reached the highest level from the first iteration of the software development process. The results produced in this step are highly dependent on the experience of an analysts. In the considered example, the team of analysts were highly experienced and therefore the results were satisfactory from the early beginning. In order to assess how application of the POP Methodology in the development process actually affects the final product quality, the next experiments should be carried out and should involve less experienced teams of analysts.

The results of the inputs completeness level are different (Fig. 2.). They show iteration to iteration progress in quality of the optimization problem definition. Four iterations were enough to identify all the required inputs.

Figure 3 illustrates the evolution of algorithms reusability level. It shows (Fig. 3) that only during the last iteration it was possible to reuse some of previously developed algorithms. It is quite obvious when we consider the requirements evolution (e.g. changing number of the required inputs). However it also presents that the application of the POP Methodology and Process Optimization Platform allows for simpler identification of possibility to reuse previously developed components. In result the total time required for final prototype development is shorter, as it has been shown in Sect. 3.1.

5 Conclusions

The POP Methodology, facilitating selection of methods of analysis, planning and development of information systems dedicated to solving optimization problems in enterprise resources management processes, proposes a complete procedure to be followed by development teams. It defines a series of steps, from the very beginning of business analysis, through the structured specification of an optimization problem, to the final delivery of a prototype solution—a software implementation of an algorithm solving the identified optimization problem. The evaluation of the POP Methodology has been performed based on a number of case studies, dedicated metrics and test scenarios executed in dedicated validation environment. The evaluation results have been presented in this paper.

The POP Methodology has been developed and evaluated in the context of the optimization needs of transportation organizations. Nevertheless, applications of the POP Methodology and the POP Platform are possible in many other domains of business activities where optimization problems can be found. The preliminary results of the POP Methodology evaluation are promising. Increased reusability level of software components and formal models has been observed. In result, the total cost of software development could be reduced by minimizing the time required for correct problem identification, description and modeling.

The future studies will be dedicated to more complex validation scenarios of software development processes and products developed with the application of the POP Methodology and the POP Platform tools. Future evaluation will include comparison of the metrics obtained during analysis of a wider set of organizations. A special interest will be put in finding the differences of quality measures between different paths defined in the POP Methodology.

Acknowledgment This work has been partially supported by the Polish Ministry of Science and Higher Education within the European Regional Development Fund, Grant No. POIG.01.03.01-02-079/12

References

1. Grzech, A., Juszczyszyn, K., Świątek, P.: Methodology and platform for business process optimization. In: Advances in Intelligent Systems and Computing—Progress in systems engineering, pp. 377–382. Springer, New York (2015)
2. Juszczyszyn, K., Kołaczek G., Świątek P., Grzech A.: Agile method of software development for resources management optimization processes. In: Information Systems Architecture and Technology: Systems Analysis Approach to the Design, Control and Decision Support. Oficyna Wydawnicza Politechniki Wrocławskiej, pp. 53–62 (2014)
3. IEEE Standard for a Software Quality Metrics Methodology. IEEE Std 1061-1998, i. doi:10.1109/IEEESTD.1998.243394 (1998)
4. Brzostowski, K., Gąsior, D., Grzech, A., Juszczyszyn, K., Kołaczek, G., Kozik, A., Rudek, R., Sławek, A., Szczurowski, L., Świątek, P.R.: Business process optimization platform for integrated information systems, Information systems architecture and technology, eds. Jerzy Świątek [i in.]. Wrocław: Oficyna Wydawnicza Politechniki Wrocławskiej, s. 13–22 (2014)
5. Jones, C.: Function points as a universal software metric. ACM SIGSOFT Softw. Eng. Notes **38**(4), 1 (2013). doi:10.1145/2492248.2492268
6. Kan, S.H.: Metrics and models in software quality engineering. Prevention (2002). Retrieved from http://portal.acm.org/citation.cfm?id=559784
7. Kaner, C., Bond, W. P.: Software engineering metrics : what do they measure and how do we know ? Direct **8**, 1–12 (2004) doi:10.1.1.1.2542
8. Vlaanderen, K., Jansen, S., Brinkkemper, S., Jaspers, E.: The agile requirements refinery: Applying SCRUM principles to software product management. Inf. Softw. Technol. **53**(1), 58–70 (2011). doi:10.1016/j.infsof.2010.08.004
9. Jones, C.: Applied software measurement: global analysis of productivity and quality. Library (vol. 38) (2008b). doi:10.1300/J111v38n01_02
10. Jones, C.: Variations in software development practices. IEEE Softw. (2003). doi:10.1109/MS.2003.1241362
11. Jones, C.: Applied software measurement. library (vol. 38) (2008a). doi:10.1036/0071502440
12. ISO/IEC 25010. Systems and software engineering—Systems and software Quality Requirements and Evaluation (SQuaRE)—System and software quality models (2011)
13. Kołaczek, G., Juszczyszyn, K., Świątek, P., Grzech, A.: Decision making in security level evaluation process of service-based applications in future internet architecture. In: Advances in Systems Science, pp. 701–714. Springer International Publishing, New York (2014)

Modeling Business Rules for Transportation Systems

Krzysztof Brzostowski

Abstract In the paper problem of business rules modelling in transportation system is considered. We focus our attention on description business domain using business rules. To this end SBVR (Semantic of Business Vocabulary and Rules) approach is considered. Based on examples of two different organizations operating in transport domain we analyze features of SBVR approach. And, at the end, it is discussed the ability of considered approach in the process of model transformation.

Keywords Model driven architecture · SBVR · Information systems · Business vocabulary

1 Introduction

One of the challenging task in development of information systems is the problem of mapping rules from business domain into specific domain of designed system. Because of business environment is highly competitive the problem of mapping rules from one domain into another is one of the most challenging task in software engineering [1, 2]. It is caused by, on the one hand, new commercial contracts and, on the other hand, customer's need.

The contracts and other documents are the basement of company strategies and policies. That it means that elaborated information system must be compliant with these documents. Moreover it must take into account customer's needs as well. It is easy to see that describing such business environment is complex task. Thus there are many different approaches that they are applied to describe business environments. The standard approach to describe the complex system is UML. It is standard language to model platform-specific description of the complex system.

K. Brzostowski (✉)
Wroclaw University of Technology, Wyb. Wyspianskiego 27, 50-370 Wroclaw, Poland
e-mail: Krzysztof.Brzostowski@pwr.edu.pl

© Springer International Publishing Switzerland 2016

J. Świątek et al. (eds.), *Information Systems Architecture and Technology: Proceedings of 36th International Conference on Information Systems Architecture and Technology – ISAT 2015 – Part III*, Advances in Intelligent Systems and Computing 431, DOI 10.1007/978-3-319-28564-1_18

The other standard is OCL which is language designed to describe expression and constraints on object-oriented models [3]. Both of these standards are designed to be use on information system domain.

Our interest is the problem of modelling business rules for information system. In this case SBVR standard is more suitable tool. SBVR is usually used in business domain to describe business requirements, specification and constraints which are coming out of company strategies and policies. To represent business constraints business rules are used [4].

SBVR standard is compatible with MDA (Model Driven Architecture). MDA is OMG's guidelines for implementation information system. MDA identifies three different models of information system [5–7]:

- Computation Independent Model (CIM) or, in different nomenclature, Business Model (BM);
- Platform Independent Model (PIM);
- Platform Specific Model (PSM).

In general, CIM or BM (business model) is provided by business analyst. This model is used to describe the organization and processes related to strategies and policies. The description should allow to execute next phase i.e. translation elaborated business model into information system. To this end it is necessary that this model must be prepared in formal representation. In order to describe organization and its environment in clear and unambiguous manner business vocabulary and business rules must be prepared. Business vocabulary is designed by domain expert to, on the one hand, to prepare business rules which are easy to compute and, on the other hand, are readable and are easy to understand by business people [5].

Having business model of the organization, the next step is to build PIM model. PIM model describes elaborated system irrespective of specific programming language, operating system or other components of information systems. It is general description of the information system, including its structures and functionalities.

The last stage is to produce specific model in which programming language, operating system etc. are taken into account.

The process of information system development requires transformation phases between discussed models [8]. In this paper we focus our attention on representation CIM/BM and its usefulness in transformation into PIM. It is one of the crucial stage of system designing because it is connecting specific business domain with IT domain. It is crucial that this transformation between business and information system ensures minimization the loss of semantic [5]. That is why CIM/BM must be defined in well and clear way.

In this paper business domain is related to transportation companies. The main activity of these companies is provide transportation services. Transportation services are the results of transportation process. The transportation process is the set of related activities to relocate some goods from starting point to destination point.

The reminder of the paper is organized as follows: after short introduction problem of business rules modelling is discussed. Subsequently model transformation methods are considered and usefulness of SBVR for this task is deliberated. In the last part of the paper SBVR's description of transportation organizations are presented.

2 Business Rules Modelling

As it was mentioned in previous section there is a need to formalize description of business domain in the process of information system development. One of the tool that can be used is previously mentioned SBVR [9–11]. SBVR is an approach helping business analyst describe business domain of organization. Benefits of SBVR are its declarative nature, representation of business rules in natural language and support for logic formulas [5, 12]. Moreover, SBVR allows to add semantic to elaborated description of the organization as well. Produced model with semantic ensures that representation is independent [13, 14]. In Fig. 1 schema illustrate process of building SBVR-based model is shown.

Business vocabulary are used to define particular business domain. The two main types of elements therein are: concepts and fact types. Concept represents business entity in considered domain. Usually, the concept can be noun concept, individual concept and verb concept [15]. Fact type are the sentences used to represent relationship between concepts.

Business rules are applied to describe the structure and behavior of the organization. To describe structure of the organization in SBVR specification definitional rules are used. On the other hand, to describe behavior of the organization behavioral rules are designed [5].

Semantic formulation in SBVR supports meaning of business rules. Usually the semantic formulations are logical rules such as: atomic formulation to specify a fact type in a rule, instantiation formulation denotes an instance of a class, logical operations like conjunction, disjunction, implication, negation, modal formulation to characterize meaning of a logical formulation [5]. The example of logical formulation of a SBVR rule is shown in Fig. 2.

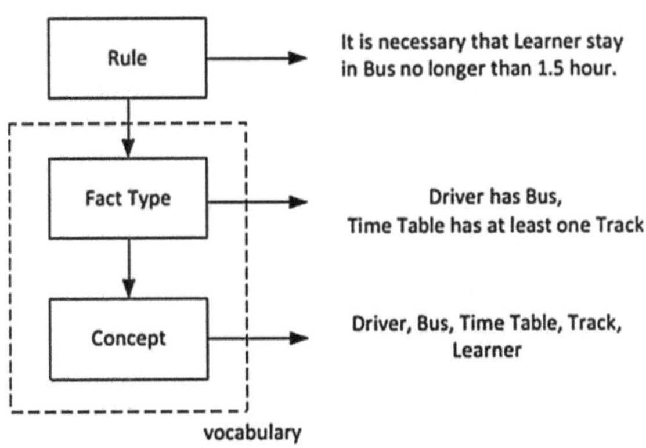

Fig. 1 SBVR-based process of model design [5]

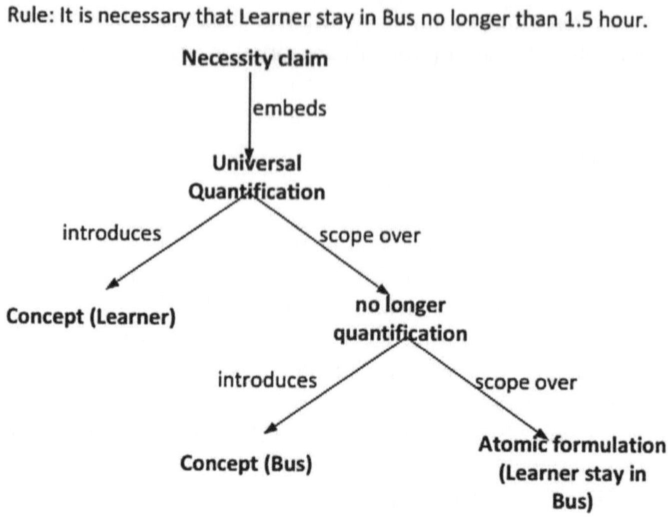

Fig. 2 Example logical formulation of SBVR rule for transportation organization

3 Methods of Model Transformation

In this section quick introduction to model transformation methods will be given. Then usefulness of SBVR for this process is considered.

The transformation of the model is the process of generation of a target model from a source model according to the transformation rules [16, 17]. Generation can be performed automatic, semi-automatic or manually.

We will focus our attention on methods to transform business model (CIM/BM) to system model (PIM).

In the literature different classification of model transformation methods are considered [18]. Taking into account type of input and output data we distinguish:

- model-to-model transformation;
- model-to-code transformation.

For model-to-model transformation the input data is the model and the output data is the model as well. For model-to-code transformation is a special case of model-to-model transformation as the input we have model and, as the result (output) we have code. Each of this category have subcategories [19].

The other taxonomy of transformation methods takes into account level of abstraction [18]:

- horizontal transformation;
- vertical transformation.

Horizontal transformation is a transformation where the source model and target model are built in at same level of abstraction. Otherwise in vertical transformation the source and target model reside at the different abstraction levels.

The another classification of transformation methods are related to previous one. In this case criterion is the number of models [18]:

- endogenous transformation;
- exogenous transformation.

Endogenous transformation characterize transformation between model represented in the same description method. Exogenous transformation is a transformation between model represented by different description method.

4 SBVR in Transportation System

In this section two models, designed according to SBVR guidelines, of two different organization from transportation domain are considered.

The first organization to be described with use of SBVR guidelines operates in domain of public transport. In the first stage vocabulary of transport domain is specified. To this end some concepts are added (Table 1). In the same table some fact types are presented. Fact type are used to define relation between two concepts. In Fig. 3 concepts and fact types are illustrated as well.

Based on specified concepts and fact types SBVR-based model of organization operating in public domain is presented in Tables 2, 3, 4, 5, 6, 7 and 8.

The second model is related to organization operates in transport domain. Similarly, in Table 9 concepts and fact types are gathered for considered organization. In Fig. 4 concepts and fact types from Table 9 are illustrated. In Tables 10, 11, 12, 13 and 14 model for the second organization with associated rules are presented.

Table 1 Concepts and fact types

Concepts	Fact types
Order	**Learner** can stay in **Bus**
Base	**Lerner** *should wait at* ***Bus Stop***
Learner	**Driver** *has* **working time**
Bus	**Order** *has* **Base**
Bus stop	**Driver** *has* **Bus**
Track	**Bus Stop** *may be included in* **Track**
Time table	**Time Table** *has at least one* **Track**
Legal working time	**Bus stop** *is kind of* **Base**
Working time	**Base** *has* **Localization**
Driver	**Working time** *should not be longer than* **legal working time**
Localization	

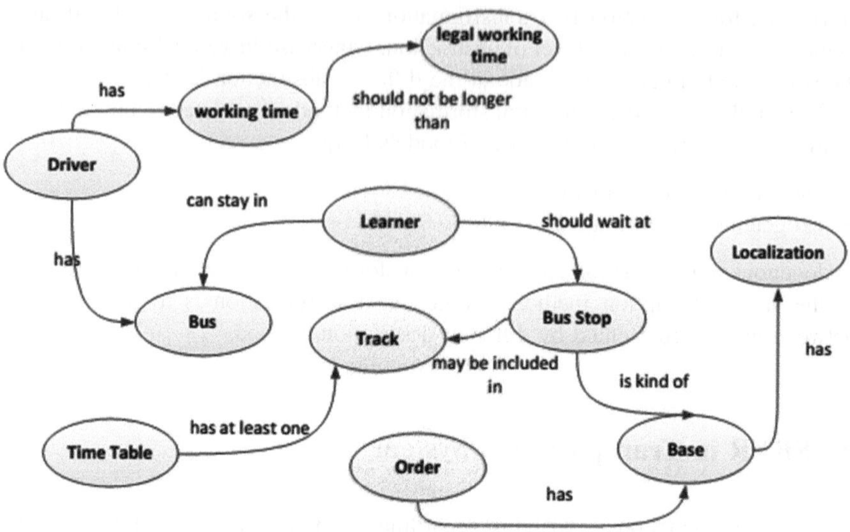

Fig. 3 SBVR concepts for the first organization

Table 2 Staying in bus

Rule: it is necessary that **Learner** can stay in **Bus** *no longer* than 1.5 h	
Fact type:	**Learner** *can stay in* **Bus**
Concepts:	**Learner, Bus**

Table 3 Waiting at bus stop

Rule: it is necessary that **Lerner** *should wait at* ***Bus stop*** at least 2.5 h	
Fact type:	**Lerner** *should wait at* ***Bus Stop***
Concepts:	**Lerner** *should wait at* ***Bus Stop***

Table 4 Driver's working time

Rule: it is obligatory that **working time** of **Driver** *should not be longer* than **legal working time**	
Fact type:	**Working time** *should not be longer than* **legal working time**
Concepts:	**Working time, Driver, legal working time**

Table 5 Order description

Rule: it is necessary that **Order** has **Base**	
Fact type:	**Order** *has* **Base**
Concepts:	**Order, Base**

Table 6 Bus is assigned to driver

Rule: it is necessary that **Driver** has **Bus**	
Fact type:	**Driver** *has* **Bus**
Concepts:	**Driver, Bus**

Table 7 Bus stops are included in track

Rule: it is necessary that **Bus Stop** may be included in **Track**	
Fact type:	**Bus Stop** *may be included in* **Track**
Concepts:	**Bus Stop, Track**

Table 8 Time table is composed of tracks

Rule: it is obligatory that **Time table** has at least one **Track**	
Fact type:	**Time table** *has at least one* **Track**
Concepts:	**Time table, Track**

Table 9 Concepts and fact types

Concepts	Fact types
Track	**Driver** *has* **Bus**
Base	**Driver** *has* **working time**
Bus stop	**Driver** *has* **Pause**
Driver	**Schedule** *has at least one* **Duty**
Bus	**Learner** *has* **Bus Stop**
Schedule	**Bus Stop** *is kind of* **Base**
Duty	**Base** *has* **Localization**
Pause	**Bus Stop** *may be included in* **Track**
Working time	**Driver** *has* **Schedule**
	Schedule *has* **Track**

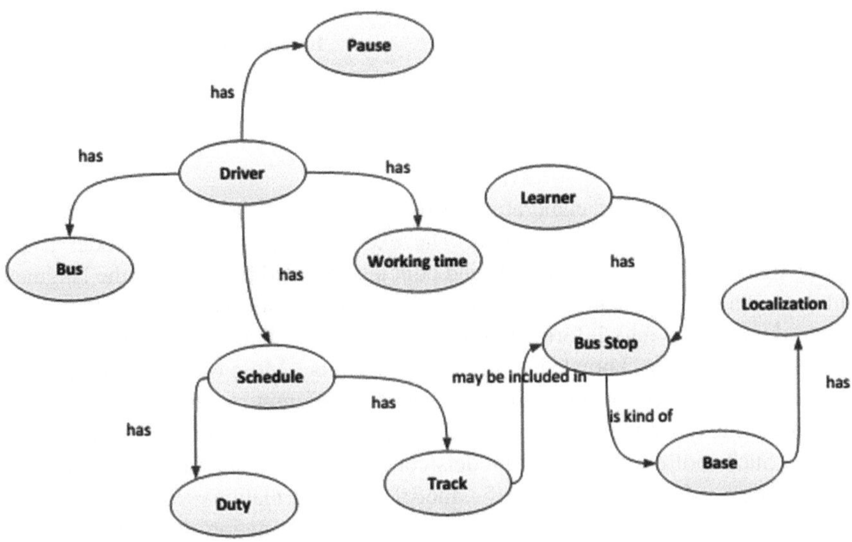

Fig. 4 SBVR concepts for the second organization

Table 10 Pause in driver's duty

Rule: it is obligatory that **Driver** has **Pause** equal 15 min working time is less than 6 h	
Fact type:	**Driver** has **Pause**
Concepts:	**Driver**, **Pause**

Table 11 Driver's work at night

Rule: it is necessary that **Driver** has **Working time** equals 10 h at night	
Fact type:	**Driver** has **Working time**
Concepts:	**Driver**, **Working time**

Table 12 Driver is assigned to bus

Rule: it is necessary that **Driver** has **Bus**	
Fact type:	**Driver** has **Bus**
Concepts:	**Driver, Bus**

Table 13 Duty in schedule

Rule: it is obligatory that **Schedule** has at least one **Duty**	
Fact type:	**Schedule** has **Duty**
Concepts:	**Schedule, Duty**

Table 14 Learner is assigned to bus stop

Rule: it is obligatory that **Learner** has **Bus Stop**	
Fact type:	**Learner** has **Bus Stop**
Concepts:	**Learner, Bus Stop**

5 Conclusion

In this work problem of elaborating business model for organization operates in transportation domain is considered. The expected feature of our model is the ability to acquire business strategies and policies provided by expert in the language similar to natural language. And, on the other hand, model should allow to transfer expert knowledge into information system. It means that model must allow to express such expert knowledge in formal manner. Moreover, modelling tools should allow to perform transformation between different models.

Taking into account mentioned requirements we decided to design our models in SBVR. Such tool allow to build models composed of business rules and business vocabularies. In viewpoint of model transformation SBVR supports model-to-model transformation, and horizontal transformation which is important in interoperability problems.

The next stage in our research is to adapt software tools to design algorithm supporting model transformation.

Acknowledgments The research presented in this paper was partially supported by the European Union within the European Regional Development Fund program Number POIG.01.03.01-02-079/12.

References

1. Feuto, P.B., et al.: Domain specific language based on the SBVR standard for expressing business rules. In: Enterprise Distributed Object Computing Conference Workshops (EDOCW) (2013)
2. Selway, M., et al.: Formalizing natural language specifications using a cognitive linguistic/configuration based approach. Inf. Syst. **54**, 191 (2015)
3. Bajwa, I.S., Bordbar, B., and Lee, M.: SBVR vs OCL: a comparative analysis of standards. In: Multitopic Conference (INMIC) (2011)
4. Bajwa, I.S., Lee, M.G.: Transformation Rules for Translating Business Rules to OCL Constraints. Modelling Foundations and Applications, pp. 132–143. Springer, Berlin (2011)
5. Raj, A., Prabhakar, T.V., Hendryx, S.: Transformation of SBVR business design to UML models. In: Proceedings of the 1st India Software Engineering Conference. ACM (2008)
6. Reynares, E., Caliusco, M.L., Galli, M.R.: A set of ontology design patterns for reengineering SBVR statements into OWL/SWRL ontologies. Expert Syst. Appl. **42**(5), 2680–2690 (2015)
7. Zur, M.M., Indulska, M.: Modeling languages for business processes and business rules: a representational analysis. Inf. Syst. **35**(4), 379–390 (2010)
8. Tantan, O.Ch., Akoka, J.: Automated transformation of Business Rules into Business Processes From SBVR to BPMN. In: 2014 International Conference on Software Engineering & Knowledge Engineering, 2014 Hyatt Regency, Vancouver, Canada (2012)
9. Hina, A., Bajwa, I.S., Bordbar, B.: SBVR2UML: a challenging transformation. Frontiers of Information Technology (FIT), IEEE (2011)
10. Cabot, J., Pau, R., Raventós, R.: From UML/OCL to SBVR specifications: a challenging transformation. Inf. Syst. **35**, 417–440 (2010)
11. Aqueo, K., Governatori, G., Sadiq, S.W.: SBVR based business contract and business rule IDE. RuleML Challenge (2010)
12. Njonko, P.B.F., El Abed W.: From natural language business requirements to executable models via SBVR. In: 2012 International Conference on Systems and Informatics (ICSAI). IEEE (2012)
13. Musham, P., Sharad S., Rashi, B.: Visual SBVR. In: Third International Conference on Digital Information Management. IEEE (2008)
14. Object Management Group (OMG).: Semantics of business vocabulary and rules specification (2006)
15. Bajwa, I.S., Lee, M.G., Bordbar, B.: SBVR business rules generation from natural language specification. In: AAAI spring symposium: AI for business agility (2011)
16. Aqueo, K., Governatori, G., Sadiq, S.: Transformation of SBVR compliant business rules to executable FCL rules. Semantic web rules, pp. 153–161. Springer, Berlin (2010)
17. Kleppe, A.G., Jos, B.W., Wim, B.: MDA Explained: the Model Driven Architecture: Practice and Promise. Addison-Wesley Professional, Boston (2003)
18. Mens, T., and P. van Gorp.: A taxonomy of model transformation and its application to graph transformation Technology. In: Proceedings of the International Workshop GraMoT (2005)
19. Czarnecki, K., Helsen, S.: Classification of model transformation approaches. In: Proceedings of the 2nd OOPSLA Workshop on Generative Techniques in the Context of the Model Driven Architecture, vol. 45, no. 3 (2003)

Author Index

© Springer International Publishing Switzerland 2016

J. Świątek et al. (eds.), *Information Systems Architecture and Technology: Proceedings of 36th International Conference on Information Systems Architecture and Technology – ISAT 2015 – Part III*, Advances in Intelligent Systems and Computing 431, DOI 10.1007/978-3-319-28564-1

Author Index

[illegible]

MIX
Papier aus verantwortungsvollen Quellen
Paper from responsible sources
FSC® C105338

If you have any concerns about our products,
you can contact us on
ProductSafety@springernature.com

In case Publisher is established outside the EU,
the EU authorized representative is:
Springer Nature Customer Service Center GmbH
Europaplatz 3, 69115 Heidelberg, Germany

Printed by Libri Plureos GmbH
in Hamburg, Germany